PROJECT APOLLO

THE MOON ODYSSEY EXPLAINED

NORMAN FERGUSON

The
History
Press

Norman Ferguson has had a lifelong interest in spaceflight and aviation, and has been writing professionally for more than fifteen years. His previous books for The History Press include *From Airbus to Zeppelin* and *The Little Book of Aviation*. He has also written comedy for the BBC and Channel 4. He lives in Edinburgh.

Photographs courtesy of NASA

Illustrations by Jemma Cox

First published 2019

The History Press
The Mill, Brimscombe Port
Stroud, Gloucestershire, GL5 2QG
www.thehistorypress.co.uk

British Library Cataloguing in Publication Data.
A catalogue record for this book is available from the British Library.

ISBN 978 0 7509 8978 7

Typesetting and origination by The History Press
Printed in Turkey by Imak

CONTENTS

INTRODUCTION

The Moon has been a source of wonderment, speculation and worship for centuries.

Its powers were seen as wide and varied: it could influence the weather, affect fertility or a person's luck. It was thought to change people's characters at different phases, and superstitions abounded. Lunar gods and goddesses were worshipped and sites built to align to the Moon.

The Moon's regular appearance in our night skies has inspired countless artists whose songs, poems, paintings, books and films reflect the mysterious, the romantic, the sinister or the comforting aspects of Luna.

Writers had speculated on what travellers would encounter when they reached Earth's only Moon; in the twentieth century humankind would finally find out.

Norman Ferguson
2019

ΛUTHOR'S NOTE

Measurements are in imperial with metric equivalents. Astronauts' names are given as normally used rather than their formal full names or ranks.

Acronyms

AGC	Apollo Guidance Computer
AGS	Abort Guidance System
ALSEP	Apollo Lunar Surface Experiments Package
ASE	Active Seismic Experiment
BOOSTER	Booster Systems Engineer
CAPCOM	Capsule Communicator
CCIG	Cold Cathode Ion Gauge
CDR	Commander
CM	Command Module
CMP	Command Module Pilot
CONTROL	Control Officer
CPLEE	Charged Particle Lunar Environment Experiment
CRD	Cosmic Ray Detector Experiment
CSM	Command and Service Module
DOI	Descent Orbit Insertion
DSKY	Display and Keyboard Unit
EASEP	Early Apollo Scientific Experiment Package

EECOM	Electrical, Environmental and Consumables Manager
EMU	Extra-Vehicular Mobility Unit
EVA	Extra-Vehicular Activity
FAO	Flight Activities Officer
FIDO	Flight Dynamics Officer
GNC	Guidance, Navigation and Control Systems Engineer
GUIDO	Guidance Officer
HFE	Heat Flow Experiment
IMU	Inertial Measurement Unit
INCO	Integrated Communications Officer
IU	Instrumentation Unit
KSC	Kennedy Space Center
LACE	Lunar Atmosphere Composition Experiment
LDD	Lunar Dust Detector
LEAM	Lunar Ejecta and Meteorites Experiment
LEC	Lunar Equipment Conveyor
LES	Launch Escape System
LLRV	Lunar Landing Research Vehicle
LLTV	Lunar Landing Training Vehicle
LM	Lunar Module
LMP	Lunar Module Pilot
LPM	Lunar Portable Magnetometer Experiment
LRRR	Lunar Ranging Retro-Reflector
LSG	Lunar Surface Gravimeter
LSM	Lunar Surface Magnetometer
LSPE	Lunar Seismic Profiling Experiment
MESA	Modularized Equipment Stowage Assembly
NASA	National Aeronautics and Space Administration
NETWORK	Network Controller
NPE	Neutron Probe Experiment
PGNCS	Primary Guidance, Navigation and Control System
PLSS	Portable Life Support System
PPK	Personal Preference Kit

PROCEDURES	Organisations and Procedures Officer
PSE	Passive Seismic Experiment
RETRO	Retrofire Officer
SCE	Signal Conditioning Equipment
SEP	Surface Electrical Properties Experiment
SIDE	Suprathermal Ion Detector Experiment
SIM	Scientific Instrument Module
SM	Service Module
SPS	Service Propulsion System
SURGEON	Flight Surgeon
SWC	Solar Wind Composition Experiment
SWS	Solar Wind Spectrometer
TELMU	Telemetry, Electrical and EVA Mobility Unit Officer
TGE	Portable Traverse Gravimeter Experiment
TLI	Trans-Lunar Injection
USAF	United States Air Force
VAB	Vehicle Assembly Building

THE MOON

Various theories have been put forward on how the Moon was formed but the most commonly accepted is the Giant Impact Theory. Around 4.5 billion years ago a large Mars-sized body collided with Earth. Most material was drawn in by Earth's gravity but the rest was collected in a process known as accretion.

The Moon, photographed on Apollo 11.

Structure

The Moon's structure is similar to Earth's in that it has an outer crust, a mantle and a core. The Moon's crust on the near side is around 43 miles (70km) thick, with the far side's double that.

Geological Features

> The Moon certainly does not possess a smooth and polished surface, but one rough and uneven, and, just like the face of the Earth itself, is everywhere full of vast protuberances, deep chasms, and sinuosities.
>
> Galileo, *Siderius Nuncius* (1610)

The terms to describe lunar features and the names of specific ones have been the subject of much discussion. Some names are officially recognised by the International Astronomical Union, founded in 1919, while others are unofficial.

Latin Term	Common Name	Examples	Notes
Mare (plural maria)	Sea	Mare Imbrium (Sea of Rains)	The lunar surface's dark flat areas, so named because early observers thought they were actual seas. Formed from volcanic lava, they cover 16 per cent of the lunar surface.
Oceanus	Sea	Oceanus Procellarum (Ocean of Storms)	Due to its status as the largest mare, the Ocean of Storms was given the distinction of being named an ocean. It is 1,600 miles (2,575km) across.
-	Basin	Orientale Basin	Large impact craters, over 186 miles (300km) wide, which contain concentric peak rings, sometimes up to six in number. All maria sit within basins.
Palus	Marsh	Palus Somnii (Marsh of Sleep)	A small plain.
Sinus	Bay	Sinus Iridum (Bay of Rainbows)	A small plain, often part of large mare.
Lacus	Lake	Lacus Timoris (Lake of Fear)	A small plain.
Terra	Highland region	Terra Vitæ (Land of Liveliness)	Brighter than the maria, the Moon's highlands make up 84 per cent of the lunar surface. The term Terra is no longer officially used.
Mons	Mountain	Mons Huygens (Mount Huygens)	Found in the lunar highlands, mountains on the Moon were formed by impacts, unlike their counterparts on Earth, which are volcanic or tectonic in origin. Mons Huygens is the highest lunar mountain, reaching a height of over 18,000ft (5,486m).

Montes	Range of mountains	Montes Apenninus (Apennine Mountains)	The Montes Apenninus range runs for over 400 miles (644km) and includes 3,000 peaks.
-	Domes	Valentine Dome	Domes are rounded, gently rising features that sometimes have summit craters. They can reach 600ft (183m) in height.
Rupes	Escarpment	Rupes Recta	These are faults in the surface or edges of craters. Rupes Recta is also known as the 'Straight Wall'.
Vallis	Valley	Vallis Schrödinger	Normally named after nearby craters.
Rima	Rille	Rima Hadley	Long and narrow depressions or channels. Some are thought to be caused by collapsed lava flows.
Promontorium	Cape	Promontorium Heraclides	These jut out into maria.
Dorsum	Ridge	Dorsum Bucher	Formed when the lava cooled and produced wrinkles. Also called veins.
Catena	Chain of craters	Catena Davy	Thought to be formed from broken-up comets.
-	Crater	Tycho	Circular depressions caused by impacts, varying enormously in size.
-	Albedo feature	Reiner Gamma	A comparatively bright lunar feature. Reiner Gamma is the only albedo feature officially identified on the near side.

Surface Composition
The Moon is covered with a layer of dust called the regolith, formed by meteorite impacts over millions of years. It can be up to 49ft (15m) deep. Amongst the rock types, igneous basalts make up most of the material found in the maria, and lighter toned anorthosites are found in the lunar highlands. Breccias were formed by rocks being fused together through meteoroid impacts.

Gardening
American scientist Harold Urey described how continual impacts erode and turn over the lunar surface as 'gardening'.

Water
Water's presence on the Moon was confirmed in 2009 when a Centaur booster was deliberately impacted in Cabeus crater and the ensuing ejecta analysed. The water, in ice form, is thought to be from a comet's impact.

Colour
Depending on different lighting conditions, astronauts observed variations in surface colour. Shades of yellow, brown, grey, white, tan and black were all seen.

Man in the Moon

There liveth none under the sunne,
that knows what to make of the man in the Moone.

John Lyly, *Endymion* (1591)

For centuries some have seen the appearance of a human face in the lunar features:

Right eye: Mare Imbrium (Sea of Rains)
Left eye: Mare Serenitatis (Sea of Serenity) and Mare Tranquillitatis (Sea of Tranquility)
Nose: Mare Vaporum (Sea of Vapours) and Mare Insularum (Sea of Islands)
Mouth: Mare Nubium (Sea of Clouds) and Mare Cognitum (Sea That Has Become Known)

Other figures believed to be visible in the Moon include: a rabbit, a hare, a man carrying sticks (sent as punishment for collecting them on a Sabbath), a woman weaving a pot, a crab, a toad, a lion, a fox, Judas Iscariot, and a hunchback sitting under a tree.

Crater Names

There are many thousands of craters and their names have accumulated over centuries. They can be named after notable scientists or polar explorers, as long as they are deceased. However, others have been rewarded: the Apollo 11 crew had craters in the Sea of Tranquility named in their honour.

Fifty craters are:

Alan   Bunsen
Alexander   Byrd
Aloha   Cassini
Amundsen   Chaucer
Anderson   Curie
Apollo   Cyrano
Archimedes   Darwin
Babbage   Einstein
Beer   Freud
Bliss   Geiger

Goddard
Grissom
H G Wells
Halley
Hippocrates
Hubble
Huxley
Ian
Icarus
Isabel
Ivan
Joy
Marco Polo
Marconi
Mary

Newton
Norman
Parkhurst
Robert
Schrödinger
Shackleton
Susan
Tereshkova
Titov
Tycho
Van de Graaff
Verne
von Braun
Wallace
Zhukovskiy

186 Miles
The largest crater is Bailly: its diameter is 186 miles (300km).

Impact Basins
The largest impact basin on the Moon's visible area is Mare Imbrium (Sea of Rains) with a diameter of 731 miles (1,160km). On the far side South Pole-Aitken Basin has a diameter of 1,550 miles (2,500km) and a depth of more than 5 miles (8km).

Zap Pits
Zap pits are small craters formed by micrometeorites.

Aristarchus
Aristarchus is the brightest crater. Although only 25 miles (40km) in diameter, it is easily seen with the naked eye due to its relative brightness in Oceanus Procellarum (Ocean of Storms).

Moonquakes
Unlike Earth, the Moon does not have tectonic plate movement. However, it is seismically active and moonquakes were detected by seismometers left behind by the Apollo missions.

Mascons
Spacecraft orbiting the Moon had variations in their orbits, put down to 'mascons', i.e. 'mass concentrations'. Mascons are believed to have been caused by large impacts bringing deeper and denser material closer to the surface. When it was thought mascons would lower Apollo 15's orbit to just 33,000ft (10km) above the lunar surface, corrective action was taken.

Magnetic
The Moon does not have a significant magnetic field but evidence it once did (around 3 billion years ago) was found in returned rocks. The lack of a north or south pole made navigation harder on the surface as compasses were inoperable.

Perigee and Apogee
The Moon's orbit around the Earth is not perfectly circular. Its furthest point is apogee and the nearest point of the oval-shaped orbit is perigee.

Distance from Earth	Position
252,900 miles (407,004km)	Apogee
221,800 miles (356,952km)	Perigee

Orbits around the Moon use the terms apolune and perilune.

Super Moon
The term 'Super Moon' describes a full Moon seen at perigee. It appears 14 per cent larger and 30 per cent brighter than when seen at apogee. A full Moon at apogee is called a Micro Moon.

59 Per Cent
Due to 'tidal locking', the Moon always presents the same face to Earth. 59 per cent of its surface can be seen from Earth, due to 'libration' – the Moon's oscillation caused by its elliptical orbit.

The Dark Side
The Moon has no 'dark side' but as almost half is never seen from Earth the concept of a 'far' or 'dark' side arose. It was first observed when Luna 3 took photographs in 1959. The first humans to see it directly were the crew of Apollo 8.

Lunar Months
The two most commonly used periods of describing how long it takes for the Moon to orbit Earth are:

Sidereal
The lunar orbital period with respect to the stars is 27.32166 days (27 days, 7 hours, 43 minutes, 12 seconds).

Synodic
The mean length of the synodic month, also known as a lunation, is 29.53059 days (29 days, 12 hours, 44 minutes, 3 seconds) and is the period between one new Moon and the next. A synodic month is longer than the sidereal as the Earth itself has moved its position relative to the Sun and it takes a few days for the Moon to 'catch up' to regain its alignment with the Sun.

Metonic Cycle
The Metonic Cycle is a period of 19 years, or 235 lunations, after which the Moon returns to exactly the same place in the sky, and its phases begin again to take place on the same day of the year.

Blue Moon

A 'Blue Moon' is the second full Moon in a calendar month or the third full Moon in an astronomical season (using equinoxes and solstices rather than calendar months) that has four full Moons.

Tides

The Moon's gravitational pull causes the water on Earth to bulge, producing two tides a day. When the Sun, Moon and Earth line up, at new and full Moons, spring tides are the result – either higher than the normal high or lower than the normal low tides. They are so named not because of the season but from them 'springing out' and then back with increased strength. Neap tides, which occur when the Sun and Moon are at right angles to each other at quarter Moons, see high and low tides experience their smallest differences.

Land Tides

The Moon also affects land and Earth's crust is raised, up to 12in (30cm). The crust sits on top of molten rock and this allows movement, although it is imperceptible to anyone standing on the affected area at the time.

Phases

New Moon	The new Moon is not seen from Earth as it is aligned with the Sun.
Waxing Crescent	Waxing is when the sunlit part of the Moon is seen, increasing each evening. Earthshine – sunlight reflecting off the Earth – illuminates the rest of the Moon.
First Quarter	Half is illuminated. The edge of the illuminated section against the unlit area is called the Terminator.
Waxing Gibbous	More than half full, comes from the Latin word *gibbus* meaning a rounded hump.

Full Moon	The whole disc can be seen, fully illuminated by sunlight.
Waning Gibbous	The amount of the illuminated Moon seen is reduced.
Last Quarter	The opposite half of First Quarter is illuminated.
Waning Crescent	The amount clearly visible reduces until only a thin crescent is seen before sunrise.

Eclipses

Solar Eclipse

Solar eclipses occur when the new Moon passes between the Sun and the Earth. It is a fortunate coincidence that the Moon and Sun seen from Earth are similar in apparent sizes: the Sun is 400 times bigger than the Moon but is 400 times farther away.

As the Moon begins to move away from totality (when no sunlight is seen), due to the rugged nature of the Moon's topography the Sun's light begins to appear in small areas first, causing a phenomenon called 'Baily's Beads' after the astronomer who described them in 1836. The final 'bead' is known as the Diamond Ring.

The last total solar eclipse to be seen in the UK was in August 1999 and the next is in 2090.

Lunar Eclipse

Lunar eclipses occur when the Moon passes into the Earth's shadow, causing it to take on a red hue. They are more common than their solar counterparts.

Annular Eclipse

An annular eclipse is when the Moon is further away from Earth than for a total eclipse. As it does not cover the whole of the Moon, it creates an effect called the 'ring of fire'.

Partial Eclipse
Only part of the Sun is obscured.

The Moon Illusion
When near to the Earth's horizon the Moon appears to be larger than its actual size. This phenomenon was discussed by notable figures such as Aristotle, Leonardo da Vinci and René Descartes and many theories have been proposed, but a conclusive explanation has yet to be agreed.

Mapping the Moon
Scientific mapping began in the seventeenth century with the advent of the telescope. Several maps were subsequently produced, including Johannes Hevelius's in *Selenographia* of 1647, and in 1651 Giovanni Riccioli defined a nomenclature system still used, with features such as Mare Tranquillitatis (Sea of Tranquility). Maps continued to be improved through subsequent years, but in the twentieth century photography took over and the use of robotic probes in the 1960s resulted in the end of Earth-based observation maps.

First Photograph
The first-ever photograph of the Moon was taken in 1840 by John Draper. Draper was a member of staff at the New York University and took the photograph on its roof. His son Henry took the first 3D images of the Moon in 1863.

'Picture of the Century'
In November 1966, Lunar Orbiter 2 took a photograph looking obliquely across Copernicus crater from an altitude of 28.4 miles (45.7km). It was the first to show real detail of the lunar topography and was described as the 'Picture of the Century' by *Life* magazine.

Picture of the Century.

Lunar Facts and Figures

1/6
Lunar gravity is one-sixth that of Earth's.

1.5
The Moon moves away from the Earth 1.5in (3.8cm) each year.

115
The maximum surface temperature is 115°C (239°F); the minimum is
−179°C (−290°F).

1,651
During the total eclipse of 21 August 2017, the Moon's shadow moved across the USA at an average speed of 1,651mph (2,657km/h).

2,159
The Moon's diameter is 2,159 miles (3,475km).

2,286
The Moon orbits the Earth at 2,286mph (3,679km/h).

35,387
The highest point on the Moon is 35,387ft (10,786m), situated close to the Engel'gardt crater on the far side. It is over 6,000ft (1,828m) higher than Mount Everest.

1,499,070
The Moon travels 1,499,070 miles (2,412,519km) in each orbit around the Earth.

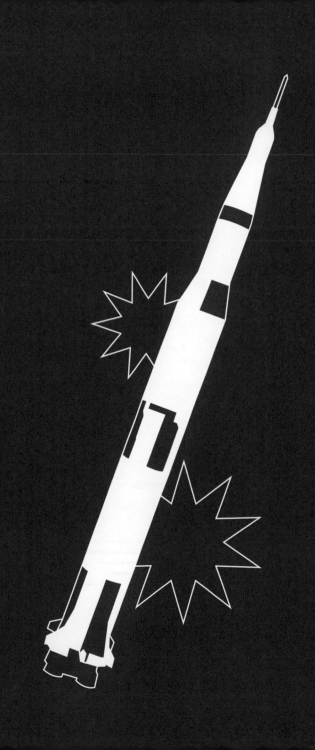

2

TO THE MOON

> This foolish idea of shooting at the Moon is an example of the absurd
> length to which vicious specialisation will carry scientists working in
> thought-tight compartments.
>
> Professor A.W. Bickerton, in 1926

Thoughts of space travel go back hundreds of years but it wasn't until
the twentieth century that scientific theories and technology were suf-
ficiently advanced to provide the means.

In Russia, a teacher called Konstantin Tsiolkovsky wrote about
how multi-stage, liquid-fuelled rockets could leave Earth orbit and be
steered in the vacuum of space. In Germany, Hermann Oberth inde-
pendently developed the concepts of staged rockets and liquid fuel,
while in America Robert Goddard was responsible for several advances
including using gyroscopes for control and mounting rocket engines on
gimbals for steering. In 1926 he successfully launched the first liquid-
fuelled rocket.

The Space Race

The Space Race began in October 1957 with the launch of Sputnik –
the world's first satellite – by the USSR. The USA had no coherent space
programme and navy, air force and army all vied for their opportunity.
A new organisation was created in 1958 to manage the space programme: NASA – the National Aeronautics and Space Administration.
A civilian-led organisation, it was tasked with putting a man into space
as part of Project Mercury. However, it was not to win this particular
race as in April 1961 the USSR sent Yuri Gagarin into orbit – another in
a line of space firsts by the Soviets.

Six Mercury missions were flown, from the 15-minute sub-orbital lob
of Alan Shepard only weeks after Gagarin's flight to the 34-hour-long
endurance flight by Gordon Cooper.

After the USSR launched its two- and three-man Voskhod flights – making the first EVA (extra-vehicular activity or 'spacewalk') in March 1965 – the USA's two-man Gemini project began. Astronauts flew in a craft similar in shape to Mercury's but with a bigger cabin and hatches able to be opened and closed in space. Ten flights proved techniques required for Apollo: rendezvous, EVA and long duration. For the latter, Jim Lovell and Frank Borman spent fourteen days in orbit in their cramped spacecraft.

Unmanned Missions

The race was also pursued with unmanned vehicles. The USSR reached the Moon first in 1959 and the same year Luna 3 sent back photographs from the far side. The USA attempted to catch up, and in 1964 Ranger 7 took America's first photographs in space. The following year Ranger 9 sent live TV footage of its final moments as it deliberately impacted the lunar surface.

In 1966 Luna 9 made the first soft landing and its successor was the first object to achieve lunar orbit. In preparation for the Apollo landings, the USA sent Surveyor craft, which achieved several landings, and Lunar Orbiter craft.

The Soviets continued to send Luna, then Zond, spacecraft, and in September 1968 Zond 5 took the first live creatures to the Moon: turtles, flies and worms.

With the Americans having already landed men in July 1969, the Soviet's Luna 16 returned soil samples to Earth via an ascent stage that landed in Kazakhstan in September 1970. Luna 17 deployed Lunokhod 1, the first lunar rover, in November 1970, which operated for 11 months, travelling 6 miles (10km).

The Soviets continued until 1976, with Luna 24 returning soil samples.

Soviet-Manned Landings

Cosmonauts were due to perform a lunar flyby as part of the Zond programme in December 1968 but technical issues cancelled the flight. For
the landing missions a cosmonaut would pilot a Lunar Craft lander to
the surface while another cosmonaut would remain in the Lunar Orbit
Module. The death of the Soviets' Chief Designer, Sergei Korolev, was a
major factor in preventing the Soviets from attempting a landing.

The N-1

The Soviet Moon rocket was to be the N-1. Its first stage was powered by thirty rockets and was only 19ft (5.8m) shorter than a Saturn V.
Despite its power, it was not successfully flown: four attempted
launches failed. The launch on 3 July 1969 saw the rocket fall back onto
the launch pad and explode – destroying the launch complex.

Wernher von Braun

During the Second World War, von Braun directed Nazi Germany's
rocket programme. As the war neared its end, his work on the V-2 bal-
listic missile made him a valuable acquisition, and von Braun and others
aimed to be captured by the Americans rather than the Soviets. They
were relocated to America where von Braun advised the US Army on its
rocket projects and was a public advocate of space travel. In 1960, von
Braun was appointed Director of the Marshall Space Flight Center. His
team was given the task of designing the Moon rocket.

A Study of Lunar Research Flights

This US Air Force secret project (also known as Project A119) investigated the possibility of exploding an atomic bomb on the Moon. It was intended that the dust cloud be seen from Earth, as a show of US strength following Russian early achievements in the Space Race. It was later revealed that the Russians had also looked at similar plans. Both projects were abandoned in favour of sending scientific expeditions.

○○○○○○○○○○○○○○○○○○○○○○○○○○○○○○○○○

Kennedy's Goal

NASA had been looking at the possibility of lunar missions since its inception but without official commitment. With Yuri Gagarin's flight, and an embarrassing military failure at the Bay of Pigs in April 1961, President John F. Kennedy was looking for an area of technological enterprise where the USA could beat its communist opponent. Kennedy discussed with advisers and then made his decision:

> I believe that this nation should commit itself to achieving the goal, before this decade is out, of landing a man on the Moon and returning him safely to Earth.
>
> Special Joint Session of Congress, Washington DC,
> 25 May 1961

This was announced just three weeks after America had put its first astronaut into space, in a flight that lasted 15 minutes. Robert Gilruth, head of NASA's Space Task Group, later revealed he reacted to the commitment by saying he was 'aghast' and not sure it would be possible to achieve.

A year later Kennedy made another speech which also served to act as inspiration for the 'great national effort' needed:

> We choose to go to the Moon in this decade and do the other things, not because they are easy, but because they are hard, because that goal will serve to organize and measure the best of our energies and skills, because that challenge is one that we're willing to accept, one we are unwilling to postpone, and one we intend to win.
>
> Rice University, Houston,
> 12 September 1962

However, in September 1963 Kennedy offered cooperation with the USSR, including a joint lunar expedition. Inside the White House there were concerns over the space programme's cost and while it is thought the USSR seriously considered Kennedy's offer, after his death in November that year, his successor Lyndon Johnson was not minded to cooperate with the Soviets.

Routes to the Moon
For centuries writers had imagined many methods of reaching the Moon: demons, swans, a large cannon, a balloon, a chariot, a copper vessel, a cloud, feathers and a whirlwind. One of them was close to the eventual reality: a firework-powered craft devised by French writer Cyrano de Bergerac in 1657.

There was much discussion amongst NASA's engineers and scientists about the best way of getting to the Moon. The first two were favoured initially:

Direct Ascent
A massive rocket would lift off, fly to the Moon, land and then return. Direct ascent would require a very powerful vehicle, with lift-off thrust of 12 million lb (5.4 million kg) – from eight F-1 rockets.

Earth-Orbit Rendezvous
Two Saturn Vs would lift off: one would carry fuel and the other the astronauts. In Earth orbit the fuel would be transferred and the refuelled rocket would then fly to the Moon.

Lunar-Orbit Rendezvous
A Saturn V would launch two spacecraft, both would fly to the Moon: one would descend to land then rendezvous with the other craft, which would bring the crew back to Earth. This method was employed, after a 2½-year campaign, by NASA engineer John Houbolt.

3

THE TECHNOLOGY

Apollo required huge advances in materials, equipment and systems. Most impressive of them all was the Moon rocket.

Saturn V
The Saturn V stood 363ft high (111m) and fully fuelled weighed 6.5 million lb (2.95 million kg). It had three stages, each discarded when its fuel was used up. The crew were located in the Command Module (CM) at the top of the 'stack'.

Stages

Stage	Engines	Fuel	Notes
S-IC	5 x F-1 rockets, each producing 1.5 million lb (680,000kg) of thrust	Liquid oxygen and kerosene	The S-IC's engines lifted the whole rocket off the launch pad and accelerated it to supersonic speed in just over a minute. They burned for around 160 seconds, burning 2,000 tons (2,032 tonnes) of fuel, enough to reach an altitude of 40 miles (64km). By engine cut-off the spacecraft was travelling over 6,000mph (9,900km/h).
S-II	5 x J-2 rockets, each producing 232,000lb (105,000kg) of thrust	Liquid oxygen and liquid hydrogen	The second stage burned for 6 minutes, enough to reach an altitude of over 110 miles (177km) and over 900 miles (1,448km) down range. Speed reached was 15,500mph (24,945km/h).

| S-IVB * | 1 x J-2 rocket producing 232,000lb (105,000kg) of thrust | Liquid oxygen and liquid hydrogen | The S-IVB propelled the spacecraft into Earth orbit at 17,500mph (28,164km/h). Unlike the previous two stages, it was not discarded immediately. After checking preparations in this 'parking orbit', the engine was re-lit to send the astronauts towards the Moon. |

* There were originally to be four stages, but the fourth stage was not renumbered.

Instrumentation Unit (IU)
Situated above the third stage, the IU housed guidance and computer systems for the initial part of the flight. Before launch the rocket's precise location in relation to its place on Earth was maintained by a theodolite positioned 689ft (210m) away which sent a beam of light through a window on the IU. After lift-off the IU aligned itself with the stars. When Apollo 12 was struck by lightning the IU kept it on course.

Saturn
The name was chosen because it was the next planet out from Jupiter and the initial Saturn designs were developed from von Braun's Jupiter rockets. At one point the space vehicle eventually called the Saturn V was going to be named 'Kronos'.

Fingerprints
The liquid oxygen propellant was so volatile that oil from a worker's fingerprint could cause an explosion. The tanks were very carefully cleaned before use.

Testing Times
During a test at the Mississippi Test Facility, five F-1 engines were fired for 15 seconds. Fifteen miles (24km) away in the town of Picayune, the sound waves shattered a bank's plate-glass window.

All-Up Testing
George Mueller, NASA's Associate Administrator for Manned Space Flight, introduced the concept of 'all-up testing'. Instead of incrementally testing one stage at a time and building up experience and confidence, the whole rocket would be flown. It was high risk. If it failed, the Apollo programme could miss its target date. If it succeeded it would save time. The results saw the third Saturn V take Apollo 8 to the Moon and the sixth launched Apollo 11. Von Braun, who had thought at least ten test flights would be needed before astronauts flew in a Saturn, wrote later, 'In retrospect it is clear that without all-up testing the first manned lunar landing could not have taken place as early as 1969.'

5,600
Jerry Lederer, Director of NASA's Office of Manned Space Flight Safety, stated that as the Saturn V had 5.6 million parts, if it achieved 99.9 per cent reliability there would still be 5,600 failures. The actual number was around thirty-five.

Launch

The launch was a spectacular event. When the engines were started, flames, smoke and steam erupted from the launch pad. As the rocket began its flight, five service arms from the launch tower disconnected and swung quickly away from the rocket.

The Saturn lifted slowly and astronauts were not always aware of the exact moment of lift-off. Those on the ground could see it rise, but it took time for the sound waves to travel. The rocket had risen silently for 6 seconds before those watching at the Kennedy Space Center (KSC) viewing area actually heard the engines. The writer Norman Mailer described it as a 'nightmare of sound' that shook the very ground he was standing on. The sound could be heard 100 miles (160km) away.

An immediate concern was colliding with the launch tower, feet away. The Apollo 8 astronauts thought the Saturn had collided, such was the vibration produced by the enormous power of the engines.

Apollo 15 lift-off.

The giant rocket took 11 seconds to reach 60mph (97km/h) but as fuel burnt off its weight reduced and, with its engines becoming more efficient in the thinner air at higher altitudes, its thrust increased, and it continued accelerating. Astronauts were pushed into their couches with a force four times that of gravity.

At 13 seconds into the flight, the rocket rolled and pitched over. It did this to gain speed and to line itself up with the proper trajectory.

During launch, if the Saturn's guidance systems failed, the commander could manually fly the rocket into orbit. Gene Cernan later said he almost dared it to fail, so he could test his flying abilities.

Launch Window

Lift-off had to take place at certain times during 'launch windows'. This was so the spacecraft would arrive at the Moon at the right time in the lunar morning with the Sun low above the horizon. The long shadows helped astronauts differentiate surface features for the landing. (It was also important to avoid the higher surface temperatures of a higher Sun.) If a lift-off time was missed, alternative landing sites could be aimed for later. Had Apollo 11's launch been delayed, *Eagle* would have landed in Sinus Medii (Bay of the Centre) or Oceanus Procellarum (Ocean of Storms).

OOOOOOOOOOOOOOOOOOOOOOOOOOOOOOOO

Saturn V Facts and Figures

1 1/4

There was enough energy produced by the Saturn's first two stages to power New York City for 1¼ hours.

4

Flames left the S-IC'S engine exhausts at four times the speed of sound.

8.9
After ignition the Saturn was held on the launch pad for 8.9 seconds until its engines reached full power, then it was released.

15
Fifteen flight-capable Saturn Vs were built, thirteen of which flew. All launched successfully.

92
92 per cent of the Saturn's weight was fuel.

211
A S-IC stage test firing produced 211 decibels (dB) of sound pressure. The current upper limit for UK workers is 140dB.

−253
The engines' liquid hydrogen was kept at a temperature of −253°C (−423°F), 20°C from absolute zero. Liquid oxygen was stored at −183°C (−297°F).

1,315
Once fully lit, the first-stage engines' exhaust reached 1,315°C (2,400°F).

2,918
On the launch pad Apollo 16 weighed 2,918 tons (2,965 tonnes) – over 300 tons more than USS *Laws*, the destroyer on which John Young served earlier in his career.

2,789
The five F-1 engines burned 2,789 gallons (12,681 litres) of fuel each second.

30,000
Thirty thousand pages of procedures were needed to prepare and launch a Saturn V.

160,000,000
The five F-1 engines produced 160 million horsepower (119 kilowatts).

Launch Escape System
At the very tip of the rocket was the Launch Escape System (LES), consisting of a protective cover for the CM and a tower containing rockets. In an emergency, these rockets would pull the CM away from the rest of the Saturn V. The LES could be used from ground level to 300,000ft (91,440m). Its rockets had twice the thrust of the Redstone that carried Alan Shepard into space in 1961.

Apollo 11 climbing.

Built in the USA

The Saturn's components were manufactured across the USA.

Component	Company	Location
Command and Service Module	North American Rockwell	Downey, California
Lunar Module	Grumman	Bethpage, Long Island, New York
Lunar Rover	Boeing / General Motors	Huntsville, Alabama
S-IC stage	Boeing	New Orleans, Louisiana
S-II stage	North American Rockwell	Seal Beach, California
S-IVB stage	McDonnell Douglas	Huntington Beach, California
S-IB rocket	Chrysler	New Orleans, Louisiana
Instrumentation Unit	IBM	Huntsville, Alabama
F-1 engine	Rocketdyne	San Fernando Valley, California
J-2 engine	Rocketdyne	San Fernando Valley, California

Logistics

These large components required well-planned travel arrangements to move them to Florida. Road transporters and barges were used, with stages built in California going through the Panama Canal. Some parts, including the S-IVB, were flown by the Super Guppy and Pregnant Guppy – specially adapted aircraft.

Command and Service Module (CSM)

	Height	Diameter
Command Module	10ft 7in (3.2m)	12ft 10in (3.9m)
Service Module	24ft 2in (7.4m)	12ft 10in (3.9m)

The CSM was composed of two parts:

Service Module (SM)
The cylindrical rear part contained fuel, water and oxygen supplies. At the rear was the Service Propulsion System (SPS) engine, used to slow the spacecraft into lunar orbit and then to propel it back towards Earth.

Command Module (CM)
The conical front part was inhabited by the astronauts during the flight. Nearing Earth on the return, the CM detached from the SM and made the re-entry and parachute-assisted landing. It had no engine but used thrusters for controlling the spacecraft's attitude.

210 Cubic Feet
The habitable volume inside the CM was 210 cubic feet (5.9 cubic metres), equivalent to three UK telephone boxes. When all three astronauts were seated wearing their spacesuits, space was so tight that their arms had to rest on top of their colleagues'.

566
The CM's control panel had 566 switches alongside seventy-one lights, forty event indicators and twenty-four instruments.

Apollo 15 CSM.

Barbecue Mode
In space the spacecraft was rotated every 20 minutes to prevent any part of it becoming too hot or too cold. Officially 'Passive Thermal Control', astronauts called it the 'barbecue mode'.

Re-entry
Re-entry to Earth's atmosphere began at a height of 400,000ft (76 miles/122km). This point was defined as Entry Interface. Encountering the atmosphere at around 25,000mph (40,000km/h) released an enormous amount of energy, mostly in the form of heat. The spacecraft had to withstand temperatures approaching 2,760°C (5,000°F). The crew were protected by a heatshield, an ablative material of phenolic epoxy resin forming the outer layer. This material would heat up and then melt, drawing heat away. Apollo 11's Mike Collins described the experience as being 'in a billion-watt light bulb'.

0.05g
After days in zero gravity, the first sign the spacecraft was returning to Earth was the indication of 0.05g on its accelerometers. This showed it was being slowed by the upper atmosphere. This was known as the 0.05g Event and took place about 30 seconds after Entry Interface. The crew would experience de-acceleration forces up to six times that of gravity.

6.5 Degrees
The CM's trajectory was ideally at 6.5 degrees above the horizontal. There was leeway of plus or minus 1 degree.

Communications Blackout
The returning spacecraft's speed in the upper atmosphere caused ionisation through which radio signals couldn't pass. This resulted in around 3 minutes of loss of signal.

Lifting Body
The CM's shape, combined with an offset centre of gravity, allowed it a small amount of lift. This was used to good effect by Apollo 11. There were storms at the designated landing site and the spacecraft extended its arrival point down range by 215 miles (346km).

Parachute Descent
Unlike the USSR's land-based landings, American spacecraft splashed down in the sea. A landing speed of 22mph (35km/h) was achieved using three 88ft-diameter (26.8m) parachutes. On Apollo 15 one of the parachutes failed after deployment but the crew still made a successful landing, albeit at a higher rate of descent.

27.5 Degrees
The CM was suspended under the parachutes at an angle of 27.5 degrees, so it did not hit the water with the full impact if fully horizontal.

Stable I and II
When the CM splashed down the parachutes were manually released. On Apollo 12, the force of the impact was such that a camera broke free of its mount and knocked out Alan Bean, whose duty it was to release the parachutes. The spacecraft was then pulled upside down and entered a position called Stable II. Stable I was when it remained the right way up.

Recovery

A flight of helicopters would fly out and navy divers would attach a sea anchor, a flotation collar and recovery raft, from which each astronaut was lifted one at a time by helicopter. The CM was lifted by crane onto the recovery ship.

Accuracy

Accuracy was important to prevent the crew from having to spend too long floating on the water in spacecraft not known for their seafaring capabilities. All missions returned to splash down in the Pacific Ocean except for Apollo 7 and 9, which landed in the Atlantic.

Apollo mission	Distance from target (miles / km)	Distance from recovery carrier (miles / km)	Aircraft carrier
7	2.2 / 3.5	8.1 / 13.0	USS *Essex*
8	1.6 / 2.6	3.0 / 4.8	USS *Yorktown*
9	3.1 / 5.0	3.5 / 5.6	USS *Guadalcanal*
10	1.5 / 2.4	3.3 / 5.4	USS *Princeton*
11	2.0 / 3.1	15 / 24	USS *Hornet*
12	2.3 / 3.7	4.5 / 7.2	USS *Hornet*
13	1.2 / 1.9	4.0 / 6.5	USS *Iwo Jima*
14	0.7 / 1.1	4.4 / 7.0	USS *New Orleans*
15	1.2 / 1.9	5.7 / 9.3	USS *Okinawa*
16	3.5 / 5.6	3.1 / 5.0	USS *Ticonderoga*
17	1.2 / 1.9	4.0 / 6.5	USS *Ticonderoga*

0.19 Per Cent

The Command Module was the only part of the spacecraft that returned to Earth. Apollo 17's CM weighed 12,120lb (5,498kg) at splashdown and, although the heaviest, it represented 0.19 per cent of the total weight launched.

CM Locations

The CMs are highly valued artefacts. All but one are located in America.

Mission	Location
Apollo 7	Frontiers of Flight Museum, Dallas, Texas
Apollo 8	Chicago Museum of Science and Industry, Chicago, Illinois
Apollo 9	San Diego Air and Space Museum, San Diego, California
Apollo 10	Science Museum, London
Apollo 11	National Air and Space Museum, Washington DC
Apollo 12	Virginia Air and Space Center, Hampton, Virginia
Apollo 13	Kansas Cosmosphere and Space Center, Hutchinson, Kansas
Apollo 14	Visitor Center, Kennedy Space Center, Florida
Apollo 15	USAF (United States Air Force) Museum, Wright-Patterson Air Force Base, Dayton, Ohio
Apollo 16	US Space and Rocket Center, Huntsville, Alabama
Apollo 17	Johnson Space Center, Houston, Texas
Apollo–Soyuz Test Project	California Science Center, California
Skylab 2	National Naval Aviation Museum, Pensacola, Florida
Skylab 3	National Air and Space Museum, Washington DC
Skylab 4	National Air and Space Museum, Washington DC

Lunar Module (LM)

	Height	Width
Ascent stage	12ft 4in (3.7m)	14ft 1in (4.3m)
Descent stage	10ft 7in (3.2m)	31ft (9.4m) (across landing legs)

The LM was unlike any craft ever flown: designed purely to fly in space with no capability of returning to Earth. It had no need of aerodynamics, and components jutted out from the main structure, which was supported by four spindly legs. Originally the 'Lunar Excursion Module', when 'Excursion' was removed from its name it was still referred to as the 'lem'.

Buzz Aldrin with *Eagle*.

Like the CSM the LM was in two parts: the descent stage had an engine to slow the spacecraft for landing and acted as the launch platform for when the ascent stage lifted off for lunar rendezvous with the CSM. The descent stage was covered with twenty-five layers of gold-coloured micrometeoroid protection and thermal insulation material.

It was not a sturdy machine. If fully loaded on Earth – and even on the Moon – the landing legs would not be able to sustain the weight. Its skin was so thin – 0.012in (0.3cm) – a dropped screwdriver would puncture it. Saving weight was critical throughout design and development: crew seats were removed and the number of windows and landing legs reduced. Astronauts were secured during zero gravity by restraint cables and Velcro panels on their boots.

Blood, Liver, Kidneys, Lungs, Skin, Eyes, Nervous System

Many parts of the human body were at risk of damage if they came into contact with the LM's propellants: 50 per cent hydrazine and 50 per cent unsymmetrical dimethyl hydrazine with the nitrogen tetroxide oxidizer. These were hypergolic: chemicals that ignited on contact. They were so corrosive they damaged the engine on contact and so engines were not tested before being used in flight.

Jettisoned

After rendezvous in lunar orbit, the ascent stages were jettisoned. Most were aimed towards the lunar surface to trigger seismometers. The exceptions were Apollo 10, which is in heliocentric orbit, and Apollo 13, which burnt up in Earth's atmosphere. Apollo 11's ascent stage was left in lunar orbit but eventually impacted the Moon.

Lunar Roving Vehicle

To allow astronauts to cover longer distances and collect more samples, vehicular transport was required. The lunar rover (officially the Lunar Roving Vehicle) went from contract signing to delivery in just seventeen months. It was carried folded up in an empty bay on the LM. The rover was powered by batteries and had four-wheel drive using steel-mesh wheels. Designed to reach 10mph (16km/h), it reached 11.2mph (18km/h) – downhill during Apollo 17.

GI Joe

General Motors personnel built a working one-sixth scale model of their proposed vehicle and used a GI Joe toy action figure (in a silver spacesuit). They caught von Braun's interest by steering it into his office by remote control.

Grover

A rover built for training purposes was nicknamed 'grover' (from 'geological rover'). It used parts from a Morris Minor. Another rover was built for astronaut driving training.

Ground Structures and Equipment

As well as spacecraft, Apollo required new equipment and structures on a scale never seen before. New launch sites were constructed at Cape Canaveral in Florida, and the main centre for administration, training, flight operations and research and development was in Houston, Texas, at the Manned Spacecraft Center. In four years it went from having 750 staff to 14,000.

Vehicle Assembly Building (VAB)

The VAB at KSC was one of the biggest buildings in the world when completed in 1966. Designed to allow several Saturn V rockets to be assembled before launch, it measures 525ft 10in (160m) high, 716ft 6in (218m) long and 518ft (158m) wide. Clouds can form inside its vast interior.

Cranes

The VAB had two bridge cranes, with a hook height of 455ft (138m), used to lift the Saturn's components. Operators had to demonstrate their skill by lowering a heavy load onto an egg without breaking it.

Apollo 16 rolls out of the VAB.

Mobile Launcher

Saturn stages were stacked on a Mobile Launcher – a 380ft-high (120m) Launch Umbilical Tower and a Launcher Base. Service arms allowed access to the rocket and carried fuel and electrical power. Some only uncoupled and swung away when sensors detected the rocket lifting off.

Launch Pads

All Moon missions lifted off from Launch Pad 39A save for Apollo 10, which used 39B. There were originally plans for three pads, as Earth-Orbit Rendezvous would have required a higher launch rate.

Crawler-Transporter

Two months before lift-off the Saturn on its launcher was moved to the launch pad. Pad 39A was 3.4 miles (5.5km) away and 39B 4.2 miles (6.8km). A unique vehicle called the Crawler-Transporter was used. This tracked machine was 131ft (40m) long and weighed 2,678 tons (2,721 tonnes) unloaded. When carrying its 5,600-ton (5,700-tonne) load it moved at 1mph (1.6km/h). Hydraulics kept the base level when going up the 5-degree ramp to the launch pad. Two Crawler-Transporters were built; they later transported the Space Shuttle.

Mobile Service Structure

This 4,600-ton (4,700-tonne), 402ft-high (123m) structure provided 'clean room' access to the spacecraft. It was moved by the Crawler-Transporter and pulled back the day before launch.

Water

Water was pumped across the launch pad and tower at a rate of 37,470 gallons (170,344 litres) a minute for fire suppression and to prevent heat damage.

Pad Survival

There were several contingencies in case of imminent explosion on the launch pad. If the astronauts were not able to use the Launch Escape System they would leave the immediate vicinity of the rocket via two options:

A high-speed lift took them (and pad crew if still on site) to a 200ft-long (61m) steel chute that descended to a circular blast room directly underneath the pad. It had oxygen and supplies for twenty people for 24 hours. An escape tunnel led 1,200ft (366m) away from the bunker.

A slide wire could carry eleven people in a basket to a bunker 2,300ft (701m) away. If required, specially prepared armoured personnel carriers were on stand-by.

Bomb

It was estimated that a Saturn V exploding was equivalent to a 0.5-kiloton (492-ton) nuclear bomb.

8

Ground personnel faced hazardous working conditions and eight workers were killed in accidents while working at KSC, including one struck by lightning while building launch pad 39B.

Red Crew

A crew of technicians were on hand to attend to any last-minute problems. One leak of liquid oxygen was fixed by wrapping wet nappies around the area. They froze and the leak was stopped.

Computing to the Moon

At the start of Apollo it was acknowledged that computers were essential to perform calculations and processing of data for all parts of the flight. At this time computers were large, occupying rooms, but to be used in spaceflight they would have to be much smaller. The MIT Instrumentation Laboratory, under Charles Draper, was given the contract to produce the Apollo Guidance Computer (AGC) in 1961. It formed part of the Primary Guidance, Navigation and Control System (PGNCS).

Its Inertial Measurement Unit (IMU) used gyroscopes and accelerators to provide a stable platform to pinpoint the spacecraft's position at all times. Despite initial plans, the on-board computer would not be the main method of navigating the spacecraft to and from the Moon. Instead it would act as a back-up to tracking from the ground. Celestial navigation using a sextant and telescope to feed information into the computer was a further back-up.

The computer calculated and controlled manoeuvring of the spacecraft, for example, the timing of engine burns to alter its orbital parameters or to adjust its attitude. The LM was also equipped with a computer to support abort manoeuvres, called the AGS (Abort Guidance System).

75

There were seventy-five programs in the spacecraft computers. Examples were P66 in the LM, used for the final landing phase, and P67 in the CM that controlled re-entry to splashdown.

Cleanlines

During testing, Draper insisted people who came back from holidays with a suntan were not allowed to work on the components in case their skin flaked.

60 Per Cent

At one point the Apollo programme was using 60 per cent of all US-made silicon chips. The AGC was one of the first computers to use integrated circuits and the procurement of so many for the space programme is credited with starting the American semi-conductor industry.

Core-Rope Fixed Memory

As well as read-write erasable memory, the computer used read-only, core-rope fixed memory, where the programs were physically tied around the computer's core. Rings and wires formed the rope. As it used a binary system, a wire through a magnetic core meant '1' and one going around meant '0'. They had to be weaved by hand in a factory and any changes were not easily made. They were used to store programs that did not need to be amended once the flight began.

LOL Method

The women at Raytheon who did the painstaking wiring were called the 'Little Old Ladies' (LOL) by staff at MIT.

Margaret Hamilton

Hamilton was the lead software developer for MIT on Apollo computers. She is credited with introducing the concept of software engineering and in 2016 received the Presidential Medal of Freedom.

Polaris

The AGC stemmed from work done on a computer that guided the Polaris ICBM.

Memory

Each computer on board the LM and the CM had 36,864 words of fixed memory, plus erasable memory of 2,048 words. Although limited by modern standards, the AGCs were efficient at the tasks assigned for each part of the flight.

DSKY

Astronauts would interact with the AGC via the 'disky' – the DSKY (Display and Keyboard Unit).

Gimbal Lock

Gimbal lock occurred when the inertial guidance system went out of alignment. The whole system had then to be carefully realigned.

37

Thirty-seven stars were used for navigational reference and three had unusual names. They were named by Apollo 1 astronauts: 'Navi' was Gus Grissom's middle name backwards, 'Regor' was Chaffee's first name backwards and 'Dnoces' was 'second' backwards, from Ed White's suffix 'II'.

Launch Control

At KSC, launches were overseen by firing rooms in the Launch Control Center. Due to the operations' complexity there were fourteen rows of consoles manned by 450 technicians, engineers and other personnel. Once the rocket had cleared the launch tower, control was handed over to Mission Control in Houston. Launch staff could then turn and view the rocket through large windows at the back of the room.

Mission Control

Every moment of each mission was overseen by Mission Control (also known as the Mission Operations Control Room) in a room equipped with large viewing screens and banks of consoles where those assigned specific positions would sit. In charge was 'Flight', the Flight Director, who had overall command of flight controllers in the room, and in back rooms where more experts were gathered. During key moments a call-and-response check of each controller was called out by Flight – a process Gene Kranz called 'going around the Horn' – to which each would respond 'Go' or 'No go'. After Apollo 11 touched down, the call was altered: 'Stay' or 'No stay'.

The Trench

The Trench was the nickname given to the front row of consoles for the Flight Dynamics controllers (RETRO, FIDO and GUIDO). The name was coined by controller John Llewellyn, as the pneumatic tube carriers around his position reminded him of the artillery shell cases in his Korean War trench.

Flight Controllers

These controllers occupied seats in Mission Control:

Abbreviation	Full Title	Area of Responsibility
RETRO	Retrofire Officer	Return to Earth and abort procedures.
FIDO	Flight Dynamics Officer	Spacecraft trajectory.
GUIDO	Guidance Officer	LM and CSM on-board navigation systems and guidance computers.
CONTROL	Control Officer	LM guidance hardware: landing radar, control thrusters, engines.
TELMU	Telemetry, Electrical and EVA Mobility Unit Officer	LM power and environmental control systems and spacesuits.
GNC	Guidance, Navigation and Control Systems Engineer	CSM guidance, navigation, and control systems.

EECOM	Electrical, Environmental and Consumables Manager	CSM electrical and life support systems.
SURGEON	Flight Surgeon	Health of astronauts.
INCO	Integrated Communications Officer	All data, voice and video communications.
CAPCOM	Capsule Communicator	Usually the only person to talk to crew during flight. Normally an astronaut.
BOOSTER	Booster Systems Engineer	Saturn V engines during launch and TLI.
PROCEDURES	Organisations and Procedures Officer	Mission procedures and rules.
FAO	Flight Activities Officer	Timeline of scheduled activities.
NETWORK	Network Controller	Ground communications between Mission Control and global tracking stations.

Bill Moon
The aptly named Moon was an EECOM during Apollo. The first person from an ethnic minority background to work in Mission Control when he began in 1964, Moon worked on Apollo 15, 16 and 17.

26
The average age of controllers during Apollo 11 was 26.

360
Each flight was controlled via mission rules, which outlined actions to be taken if certain situations arose. For Apollo 11 they amounted to 360 pages.

Communications and Tracking

Astronauts relied on constant communication between them and Mission Control. Voice, TV and telemetry as well as tracking information were carried on the S-band range of frequencies (2–4GHz). A network of Earth-based stations was used. Three primary stations were equipped with 85ft-diameter (26m) antennae able to carry the signals at long distance. They were located around the world, to provide consistent coverage, at:

- ❑ Goldstone, California, USA
- ❑ Honeysuckle Creek, Australia
- ❑ Madrid, Spain.

Other ground stations, ships and aircraft ensured coverage at different stages of the mission. The 210ft (64m) dish at Parkes in Australia received Apollo 11's TV pictures from the Moon, despite it being operated in strong winds beyond its safety limits.

Beep

Voice communications were punctuated by a short 'beep'. This was added deliberately to indicate to the radio network when speech was beginning and ending.

GALS

The scale of the project required large administrative support. Amongst those were the secretaries assigned to the astronauts. One group who continued to meet up after leaving NASA were the 'GALS' – Girls of Apollo Launches.

Spacesuits

Apollo spacesuits had different requirements from those in Mercury and Gemini – they had to protect and sustain those working in the extreme lunar temperatures and also provide protection against micrometeroid impact.

Apollo suits were officially the A7L Extra-Vehicular Mobility Unit (EMU). Each consisted of twenty-one layers of material, such as nylon, Nomex, Teflon and Mylar, and contained over 500 parts. Underneath, the astronauts wore a long-john-type, cotton-knit Constant Wear Garment, and underneath this a belt of bio-medical sensors, which provided data on heart rates and respiration. During EVAs astronauts would wear a Liquid Cooling Garment that ran cool water through small tubes around the body. This followed lessons learnt in Gemini, when astronauts experienced overheating.

Spacesuits were worn for launch and any critical part of the mission, such as undocking and landing. Part of the suit was the polycarbonate 'fishbowl' helmet. For surface EVAs, suits were supplemented with over-boots, gloves, a protective helmet assembly (with a gold-plated visor to protect against heat and light) and a backpack called the Portable Life Support System (PLSS), which provided oxygen, water for cooling and communications equipment. On Earth the suit with PLSS weighed over 180lb (82kg).

When out of their spacesuits, astronauts wore a two-piece Teflon Inflight Coverall Garment, although on some flights it was warm enough just to wear their long-john underwear.

Valsalva Device

With the helmet on it was impossible for astronauts to equalise pressure in their ears by blowing out while holding their nose closed. An attachment called the Valsalva Device was placed inside the helmet to allow this procedure. It also served as a useful place to scratch an itchy nose.

Playtex
Spacesuits were manufactured by a division of the company that made women's underwear under the brand name Playtex.

Snoopy
Underneath their helmets astronauts wore a Communications Carrier, similar in appearance to flying helmets worn by early aviators. It contained earphones and microphones and was called the 'Snoopy cap' because of its black and white material, similar to the cartoon character. Snoopy became the symbol of NASA's quality assurance programme introduced after the Apollo 1 fire, and silver Snoopy pins are still awarded to employees for outstanding contributions to flight safety.

Checklists
As it was impossible to remember all the actions required for each activity, comprehensive checklists were prepared. Astronauts carried these checklists while on the lunar surface, and in the famous 'Man on the Moon' photograph, Buzz Aldrin's left wrist is angled upwards as he reads off the next activity. On Apollo 12, Bean and Conrad's checklists contained unofficial additions: nude photographs from *Playboy*.

Cuff Checklist
For surface EVAs, checklists gave the timeline in minutes from the start, then brief details of each activity.

Pete Conrad's EVA 1 checklist, Apollo 12

Timeline	Details	Notes
0 +10	CDR EGRESS DEPLOY LEC AND MESA	Conrad (CDR) is to leave the LM, descend the ladder, then deploy the LEC (Lunar Equipment Conveyor) – a pulley to move heavy or bulky items to or from the surface. The MESA (Modularized Equipment Stowage Assembly) is in a compartment on the side of the LM. It contains the TV camera, geology equipment (hammers, scoops, etc) and sample return containers.
0 +18	FAM MOBILITY & STABILITY CG SHIFT-FORWARD, BACK, SIDE DOWNWARD REACH ARM MOTION WALKING (BALANCING, BOOT PENETRATION, TRACTION, SOIL SCATTERING AND ADHESION)	Conrad would have 5 minutes to familiarise himself with moving on the lunar surface in the one-sixth gravity. This page also had the addition of a cartoon astronaut bouncing on the Moon.
0 + 23	CONTINGENCY SAMPLE	Taking an amount of lunar soil in case the crew had to make a rapid departure.

Time Pressure

Each EVA was always against the clock, with events often taking longer than planned. Apollo 14's Ed Mitchell described the workload as being '120 per cent of human capacity'. As each sample was photographed to record its location and then placed in numbered bags, it took time.

Tindallgrams

Bill Tindall, Chief of Apollo Data Priority Coordination, was known for producing memos known as 'Tindallgrams' in which he summarised meetings or outlined changes to procedures. One ended, 'They [decisions on rendezvous] will only be changed if there is <u>a darned good reason</u> – not just to make things a little better!' Flight Director Gene Kranz gave Tindall a seat next to him during Apollo 11's landing as a reward for all his work on refining the landing techniques.

4

THE
ASTRONAUTS

The first astronauts were chosen primarily for their test-piloting experience. Climbers, matadors and scuba divers were considered but test pilots were selected because they were used to high-risk operations and could follow orders. The term 'The Right Stuff' was devised by American writer Tom Wolfe to sum up the qualities of the confident and capable test pilot, and some of the test pilot brethren thought the early astronaut volunteers were nothing more than 'spam in the can', as the Mercury capsules' automation and ground control left little for a pilot to do. This thinking receded when the complexity of the subsequent spacecraft and associated astronaut input became clear.

Later astronauts did not have to be test pilots but did require an advanced degree. Pressure from the scientific community meant that non-pilots who had a science background were also accepted.

Apollo astronauts came from several intakes:

'Original Seven' (1959)
Scott Carpenter
Gordon Cooper
John Glenn
Gus Grissom
Wally Schirra
Alan Shepard
Deke Slayton

'The New Nine' (1962)
Neil Armstrong
Frank Borman
Pete Conrad
Jim Lovell
Jim McDivitt
Elliott See
Tom Stafford

Ed White
John Young

'The Fourteen' (1963)
Buzz Aldrin
Bill Anders
Charlie Bassett
Alan Bean
Gene Cernan
Roger Chaffee
Mike Collins
Walt Cunningham
Donn Eisele
Ted Freeman
Dick Gordon
Rusty Schweickart
Dave Scott
C.C. Williams

'The Scientists' (1965)
Owen Garriott
Edward Gibson
Duane Graveline
Joe Kerwin
Curt Michel
Harrison Schmitt

'The Original Nineteen' (1966)
(This group were given their
tongue-in-cheek nickname by
John Young)
Vance Brand
John Bull
Gerald Carr
Charlie Duke

Joe Engle
Ron Evans
Edward Givens
Fred Haise
Jim Irwin
Don Lind
Jack Lousma
Ken Mattingly
Bruce McCandless
Edgar Mitchell
Bill Pogue
Stuart Roosa
Jack Swigert
Paul Weitz
Al Worden

Two other groups were added during Apollo. In 1967 the 'XS-11' (the Excess Eleven's self-deprecating nickname reflecting their view of how crucial a role they would play) and 1969's Group 7 were USAF personnel transferred when the Manned Orbiting Laboratory project was cancelled. None of these would fly to the Moon but several would fly on the Space Shuttle.

First-Born
Of the twenty-nine astronauts who flew on Apollo, twenty-two were the first-born amongst their siblings or were an only child.

Technical Assignments
As well as their training and other commitments, astronauts were given areas of technical responsibility where they liaised with NASA personnel and equipment and spacecraft manufacturers. For example, Neil Armstrong worked on simulators and Mike Collins on spacesuits.

Crew Selection
Deke Slayton was Director of Flight Crew Operations and the main person who chose who would fly in each crew. The reasons behind his selections were not made public and much would depend on how he viewed a particular astronaut.

Back-Up Crew
Each mission's prime crew had a back-up crew (and also a support crew) assigned. Several back-ups were used, such as Jack Swigert for Apollo 13, called into action only days before launch. One of the back-up crew's tasks was to organise 'pin parties'. These were held to mark the crews' return from space and to award first-time space flyers their gold astronaut pin.

Training
Astronauts spent many hours training, much of it in simulators replicating CM and LM cockpits. Training was intensive, with every aspect from launch to re-entry being practised over and over again, so much so that some astronauts felt almost anti-climatic when they reached the Moon.

2,000
Apollo 11 astronauts spent 2,000 hours in the simulators in the seven months before their flight.

'The Great Train Wreck'
John Young called the CM simulator the Great Train Wreck due to its appearance – sections jutted out at all angles.

LLTV
Flying machines called the Lunar Landing Research Vehicle (LLRV) and then a development called the Lunar Landing Training Vehicle (LLTV)

– unofficially the 'flying bedstead' – were built to simulate the LM's flight characteristics. In May 1968 Neil Armstrong was forced to eject seconds before his LLRV went out of control and crashed. Despite this, he valued the training they provided.

Gravity

All astronauts faced operating in zero gravity, and those landing would experience one-sixth gravity, and so methods were devised to give experience of these conditions. 'Vomit comet' aircraft flying parabolic arcs would give brief periods of weightlessness, as would training in swimming pools. Hoists suspended from moving trucks were used to replicate lunar gravity and another method tried was to suspend the astronaut from a rig while he was 'standing' on his side.

T-38

Astronauts were keen to maintain their proficiency in flying fast jets and a fleet of T-38 supersonic aircraft were kept by NASA for this purpose. They also served as vehicles for the many cross-country flights made to the various manufacturing centres across the USA. These high-performance aircraft also helped the astronauts maintain tolerance to high-g manoeuvres and helped alleviate space sickness.

Survival Training

In case emergency landings were made away from the sea, training was given in Panama on jungle survival and in Nevada and Washington state on desert survival.

Moon on Earth

In order to gain the maximum results possible, astronauts were trained in geology in the classroom and in the field. Geologists such as Professor Lee Silver from Caltech and Gordon Swann from the US Geological Survey were brought in. Not all astronauts were enthusiastic about the subject but some, such as Dave Scott, relished the opportunity.

Geology field trips were organised, predominantly in the USA although Canada, Iceland, Germany and Mexico were visited. Places in the USA they visited included:

- ❏ Big Bend, Texas
- ❏ Black Canyon Crater Field, Arizona
- ❏ Cinder Lake Crater Field, Arizona
- ❏ Craters of the Moon, Idaho
- ❏ Grand Canyon, Arizona
- ❏ Big Island, Hawaii
- ❏ Valley of Ten Thousand Smokes, Alaska
- ❏ Meteor Crater, Arizona
- ❏ Nevada Test Site, Nevada
- ❏ Orocopia Mountains, California.

Astronauts also trained at sites created to mimic lunar conditions. At KSC a sand pile was made; however, it occasionally encountered conditions unlikely to be replicated on the Moon: the craters filled with water at high tides. In Arizona craters were created by triggered explosions to exactly replicate landing sites.

Man in the Moon

Eugene Shoemaker was an astrogeologist who had a large influence on the geological component of Apollo. A health issue prevented him becoming an astronaut but he helped train those selected. After his death, Shoemaker's ashes were sent to the Moon via Lunar Prospector, and on 31 July 1999 he became the first person to have his remains interred on another celestial body.

Life

In 1959 NASA signed a contract with *Life* magazine, providing the periodical with officially sanctioned and exclusive stories on the astronauts and their lives, wives and families. Each of the Mercury astronauts received an equal share of the $500,000 contract (approximately $25,000 a year). As it was difficult to secure elsewhere, $100,000 of life insurance was also included. There was criticism that the contract was unfair and restrictive – the stories *Life* published were never critical – and it was almost cancelled in 1962. It was extended but with the growing astronaut numbers, the amount received by subsequent astronauts reduced.

Salary

Whether civil or serving military pilots, astronauts were on government pay scales. Neil Armstrong was technically a civil servant and while on Apollo 11 was on pay grade GS-16, earning $30,054, while crewmates Buzz Aldrin received $18,622 and Mike Collins $17,147 a year.

Proud, Thrilled, Happy

Astronauts' wives faced life with husbands who were very often away from the family home. They had to manage the household, care for their children and face the press during stressful periods. The wives often used a standard response when asked how they felt about their husbands' exploits in space: they were 'proud, thrilled and happy'. After Apollo 12, Sue Bean, Barbara Gordon and Jane Conrad held up cards adorned with these three words.

Moon Walkers

Twelve astronauts have walked on the Moon:

Apollo 11

Neil Armstrong (1930–2012)

Armstrong's keen interest in flying saw him join the US Navy where he flew seventy-eight missions in the Korean War. He became a civilian test pilot, flying the X-15 high-speed research vehicle before joining the space programme. During Gemini 8 he had to abort when the spacecraft began spinning out of control following the first-ever docking of two vehicles. After Apollo 11 he left NASA to become a university professor and then followed business interests. Armstrong was a private man who did not seek publicity and always regarded his own fame as being undeserved and diverting from the efforts of everyone involved.

Neil Armstrong suits up for Apollo 11.

Buzz Aldrin (1930–Present)

Aldrin flew F-86 Sabres in the Korean War, shooting down two aircraft. Known as 'Dr Rendezvous', he produced valuable research findings on orbital mechanics utilised by Gemini and Apollo. Aldrin flew on Gemini 12 and carried out EVAs totalling 5½ hours. After Apollo 11 he commanded the US Air Force's Test Pilot School. Aldrin remains an enthusiastic advocate of spaceflight, especially towards Mars exploration.

Apollo 12

Pete Conrad (1930–99)

Conrad was a Navy test pilot before joining NASA and flew on Gemini 5 before commanding Gemini 11. He remained in NASA after Apollo 12 and commanded Skylab 2 before working for aircraft manufacturer McDonnell Douglas. Conrad showed his renowned sense of humour by returning a blank sheet of paper during a psychology test saying it was 'upside down'.

Alan Bean (1932–2018)

Bean had been instructed by his future Apollo commander Pete Conrad while in the Navy. He became an astronaut in 1963 and flew in 1969, becoming the fourth man to walk on the Moon. Like Conrad, Bean commanded a Skylab mission. When he left NASA he devoted himself to capturing on canvas his experiences on the Moon.

Apollo 14

Alan Shepard (1923–98)

Shepard served in the US Navy during Second World War and following the war's end began flying training. After a period as a test pilot he became one of the 'Mercury Seven' and was the first American in space in 1961. He was grounded with an inner ear medical problem and became Chief of the Astronaut Office. Surgery corrected his problem and he commanded Apollo 14. While still an astronaut, Shepard made money through banking and property investment.

Edgar Mitchell (1930–2016)

Mitchell joined NASA in 1966 and his only space flight was Apollo 14. During the flight he experienced a feeling of universal connectedness and after returning to Earth spent time investigating the nature of consciousness, forming the Institute of Noetic Sciences in 1973.

Apollo 15

Dave Scott (1932–Present)

Scott flew fighter jets for the US Air Force before becoming a test pilot. He joined the astronaut corps in 1963 and three years later went into space on Gemini 8. Scott flew on Apollo 9 and commanded Apollo 15. He remained in NASA for the Apollo–Soyuz project and then worked in business.

Jim Irwin (1930–91)

Irwin was a US Air Force test pilot, flying the Mach 3 YF-12 Blackbird. He flew in space once, on Apollo 15, which had a huge effect him; he devoted the rest of his life to religion. In the 1970s and 1980s he searched for the site of Noah's Ark on Turkey's Mount Ararat.

Apollo 16

John Young (1930–2017)

As a test pilot Young set time-to-altitude records in the F-4 Phantom. He joined NASA in 1962 and flew on Gemini 3 then commanded Gemini 10. Young was a member of a select club (along with Gene Cernan and Jim Lovell) who flew twice to the Moon. He travelled into space two more times, in the Space Shuttle, including its first flight in 1981. When he retired in 2004 Young had spent 42 years as part of America's space programme.

Charlie Duke (1935–Present)

Duke's flying career began with the US Air Force, where he flew fighters. He was a test pilot instructor before becoming an astronaut. Duke was Capcom for Apollo 11's landing and then flew on Apollo 16. He left NASA in 1975 and furthered business interests. As with Jim Irwin, Duke's journey to the Moon also furthered a keen desire to devote energy to religion.

Apollo 17

Gene Cernan (1934–2017)

Cernan flew fighters for the US Navy before joining the space programme. He first flew into space in 1966 on Gemini 9, where he became the second American to carry out a spacewalk. He was LMP on Apollo 10 before commanding Apollo 17. Cernan remained a passionate advocate of the potential of manned spaceflight.

Harrison 'Jack' Schmitt (1935–Present)

Schmitt's geology background helped familiarise Apollo astronauts with the subject. When the later Apollo flights were cancelled he replaced Joe Engle on the last lunar landing mission. The only geologist-astronaut to reach the Moon remained with NASA until 1975 when he entered politics, becoming a senator for New Mexico.

Fallen Astronauts

Eight US astronauts died before they could reach space as part of Apollo:

Astronaut	Cause	Date
Ted Freeman	aircraft crash	31 October 1964
Charlie Bassett	aircraft crash	28 February 1966
Elliot See	aircraft crash	28 February 1966
Roger Chaffee	launch pad fire	27 January 1967
Ed White	launch pad fire	27 January 1967
Gus Grissom	launch pad fire	27 January 1967
Edward Givens	car accident	6 June 1967
C.C. Williams	aircraft crash	5 October 1967

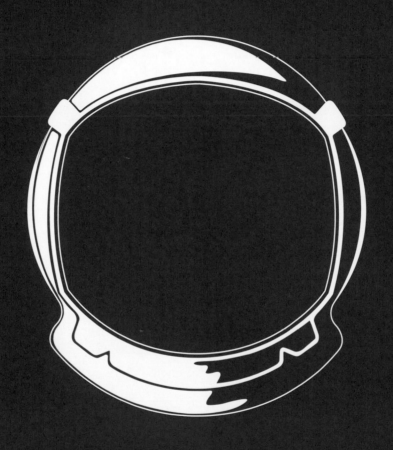

5

LIFE IN SPACE

Astronauts would spend extended time in space and provision had to be made so they could function effectively.

Pre-Launch Schedule

Astronaut schedules were filled with training, procedure checking, travel to external sites, media opportunities and keeping fit. Long hours were the norm. Little time was left to spend with families. Al Worden spent a year and a half working at North American in California, commuting every Sunday night and returning the following Friday. Strains were inevitable and divorces not uncommon. Eight of the eleven Moonwalkers who were married at the time of their missions later divorced.

> If you think going to the Moon is hard, try staying at home.
>
> Barbara Cernan

The pace intensified for crews next to fly but, in the final days before lift-off, the pace slackened as it was felt a tired crew would not be able to perform effectively. Apollo 11's crew spent pre-launch day on the beach relaxing and Buzz Aldrin went metal detecting.

Medical isolation was imposed three weeks before lift-off to prevent infection and the flight crew were restricted to training areas and living quarters.

Oxygen

Astronauts breathed pure oxygen for several hours before launch to clear their blood of nitrogen. Without doing so, the reduction in pressure as they climbed could lead to 'the bends'. They carried their own oxygen supply to the launch pad in portable ventilators.

Radiation

A risk the crews faced was that of radiation. In 1958 scientist James Van Allen found evidence of radioactive areas around the Earth. These Van Allen Belts, containing charged particles caught in Earth's magnetic field, presented problems to astronauts travelling further than 350 miles (563km) from Earth. NASA mission planners considered the spacecraft would provide enough protection, as they passed through the belts in a period of several hours.

Food and Drink

Providing sufficient and tasty food without access to traditional kitchen facilities was a challenge. How to balance nutritional needs against technical demands on water and electricity, as well as storage and time to prepare, were all considered in great detail.

Mercury and Gemini had used food squeezed out of tubes or small finger food. For Apollo, rehydrating food with the plentiful by-product of the fuel cell system was done and hot water was provided. Hydrogen bubbles were not completely removed from the water and stomach gas was the result. Buzz Aldrin later joked that the crew could have done the job of the spacecraft's thrusters.

Food provision developed and reliance on rehydrating food was reduced. Some came in 'wetpacks' and could be eaten with a spoon and 'intermediate moisture bites', such as fruitcake and brownies, were also included.

John Young and Gus Grissom ran into trouble after a corned beef sandwich was smuggled on board their Gemini flight, but sandwiches became part of the Apollo menu. The bread was subjected to 50,000rads of cobalt-60 gamma irradiation to pasteurize it.

It was intended that 2,300 kilocalories of food was provided each day; astronauts could select their own items before a flight. The menus for Apollo included:

❏   Bacon squares
❏   Sausage patties
❏   Cinnamon toast bread cubes
❏   Peaches
❏   Cornflakes
❏   Frankfurters
❏   Chicken soup
❏   Turkey and gravy
❏   Ham and potatoes
❏   Chicken and rice
❏   Spaghetti
❏   Tuna salad
❏   Banana pudding
❏   Butterscotch pudding
❏   Chocolate cubes
❏   Jellied fruit candy
❏   Coffee
❏   Orange drink

Astronauts required extra sustenance while working during long-duration EVAs, and food bars and water or fruit juice were provided inside the neck ring of their spacesuits.

Toilet

One of the most frequently asked questions for astronauts was how they went to the toilet. In the CM, urine was collected before being dumped into space. It would immediately freeze and form crystals dubbed 'Constellation Urion' by Wally Schirra. When astronauts were wearing their spacesuits urine was dumped once back inside the spacecraft.

Defecating was a more complicated matter that involved each astronaut having to remove their coveralls and put on a special faecal collection bag, taped to the buttocks. Once finished, the bags, containing a germicide to prevent bacteria and gas forming, were tied up. It could take up to an hour to perform the task. Despite the odour and risk of leakage, the bags were brought back for analysis. Astronauts would try to avoid or postpone it, and on the eleven-day Apollo 7 mission the entire three-man crew only carried out twelve bowel movements. As the faecal bag method couldn't be used during EVAs, absorbent shorts were provided.

Shaving

Astronauts on early space flights were not able to shave. On Apollo electric shavers were tried but the most effective method was the traditional wet shave. Gene Cernan described it as being 'reborn' but crew choices varied: Apollo 15's crew returned with beards and Mike Collins sported a moustache.

Washing

With no showers available, cleaning was done with damp cloths. Jim Irwin took scented soap and said the smell helped them think they were 'almost clean'.

Sleep

The amount of sleep enjoyed varied. On Apollos 7 and 8 it was difficult to sleep as one crewmember was awake on watch duties at all times. For later missions the crew all slept at the same time. To prevent them bumping into the CM's controls while asleep in the weightless environment, restraints were provided.

On the LM, arrangements were initially rudimentary. Buzz Aldrin slept curled up on the LM's floor and Neil Armstrong was propped up on the engine cover with his legs in a makeshift sling. Despite being awake for over 21 hours they both struggled to get enough rest. Both were wearing their spacesuits and Armstrong was not helped by the light from the Earth shining directly through the LM's telescope into his eyes. It was also cold inside the cabin.

LM astronauts were later provided with hammocks, and for the J-missions suits were taken off after EVAs. Adequate rest could be obtained, despite the noise of the lander's pumps and fans, which Jim Irwin described as like 'sleeping in a boiler room'. Another factor was the altitude of the LM. Apollo 14's *Antares* had a tilt of 7 degrees and sleep was interrupted by thoughts it might tip over.

Although sleep was essential, the astronauts were not necessarily keen. On Apollo 8, Borman's orders to rest were resisted by Bill Anders who wished to make the most of the time in lunar orbit.

Music

Personally chosen music could be played on cassette recorders. On Apollo 10 the music included appropriate songs by Frank Sinatra such as 'Fly Me to the Moon', 'Come Fly With Me' and 'Moonlight Serenade'. On Apollo 12 The Archies' pop song 'Sugar, Sugar' resulted in zero-g dancing. At the other end of the musical spectrum Ken Mattingly on Apollo 16 listened to classical music, including the soundtrack to *2001: A Space Odyssey*. However, the choices were not always to everyone's taste: on Apollo 9 Dave Scott hid Rusty Schweickarts's tape of classical music until near the end of the mission.

Claustrophobia

Astronauts were used to operating in small aircraft cockpits and claustrophobia was not often experienced. However, during his development work on the spacesuits, Mike Collins had several episodes, which led to him considering leaving the programme. He was able to overcome the panicky feelings and continued.

Fear

Fear was not something astronauts admitted to but Alan Bean stated later he was one of the more fearful astronauts and John Young described how when seeing the Earth reduce from a globe to a flat disc, he had the feeling they might have 'bitten off more than we could chew'.

6

THE APOLLO MISSIONS

The Apollo programme was always under intense and public pressure to achieve Kennedy's goal. Missions were incremental: each had to be completed successfully before an attempt on the next type could be made.

Apollo Mission Types

Type	Details	Mission
A	Unmanned Saturn V and CM re-entry test	4, 6
B	Unmanned LM test	5
C	CSM testing in low-Earth orbit	7
C prime	Lunar orbital	8
D	CSM and LM testing in low-Earth orbit	9
E	CSM and LM testing in high-Earth orbit	Not flown
F	CSM and LM testing in lunar orbit	10
G	First lunar landing	11
H	Precision landing with two lunar EVAs	12, 14
I	CSM lunar orbital survey	Not flown
J	Extended duration lunar landing, with rover and three EVAs	15, 16, 17

The Crew

Each Apollo mission had three crew members, with specific roles:

Title	Abbreviation	Details
Commander	CDR	Overall spacecraft command; piloted LM on landing.
Command Module Pilot	CMP	In charge of navigation, flew CM in lunar orbit and on re-entry.
Lunar Module Pilot	LMP	Monitored systems while CDR operated LM flight controls.

APOLLO 1

Crew:	Gus Grissom (Command Pilot) Ed White (Senior Pilot) Roger Chaffee (Pilot)
Mission patch:	Astronauts' surnames surround a CM flying over USA as the Moon rises over the Earth's horizon
Date (launch):	21 February 1967 (scheduled)
Date (return):	n/a
Mission duration:	Up to two weeks (scheduled)
LM:	n/a
CM:	Unnamed
Landing site:	n/a
Time on the Moon:	n/a
Mission objective:	To make the first manned flight of CM and test the spacecraft in Earth orbit

Block 1
The CM was an early version, designed at the start of the programme. Lunar flights would use the Block II.

Lemon
Problems were constantly found. Grissom became so irritated he put a lemon on top of the CM simulator and the crew were photographed praying in front of a model of the spacecraft.

Plugs Out Integrated Test
On 27 January all systems were to be tested, with the crew in the spacecraft on the launch pad. 'Plugs out' meant the spacecraft ran on its own power, simulating a countdown. The Saturn IB on which the CM sat was unfuelled and the test was not regarded as being hazardous.

How are we going to get to the Moon if we can't talk between three buildings?

Minutes before the fire, Grissom expressed his frustration at communications issues. The test was halted to allow attempts to resolve the problems.

We've got a fire in the cockpit.

Chaffee's announcement indicated an urgent problem. Those watching on CCTV saw flames inside the cabin. Technicians rushed to help but were forced back as the craft ruptured, billowing out smoke and flames. Efforts to rescue the crew continued despite the choking smoke. There was an added risk: the Launch Escape System's rocket might be triggered.

15 Seconds

An egress exercise had been scheduled at the end of the test. A minimum of 70 seconds was needed for the crew to escape unassisted. There were three hatches to remove and the inward-opening inner one, weighing 55lb (25kg), was released by a crew member turning six latch assemblies.

The pressure inside the CM was 16.7psi; the exterior air pressure was 14psi. A valve was fitted to reduce cabin pressure but it was rendered useless as the heat caused internal pressure to rise rapidly; the fuselage ruptured with pressure having reached 29psi. It was 15 seconds from the first verbal alert to the end of all transmissions from the astronauts.

Ironically, the hatch was designed to counter the problem Grissom encountered after his Mercury flight when the outward-opening hatch blew open and the spacecraft sank.

Investigation

> Deficiencies existed in Command Module design, workmanship and quality control.
>
> Report of Apollo 204 Review Board,
> 5 April 1967

When technicians were able to open the hatch they found all three dead. An official investigation found that a spark from exposed wiring was probably the main cause of the fire. It was compounded by leaking pipes containing combustible coolant and there was too much flammable material in the cockpit: Velcro and nylon netting had ignited in the oxygen-rich environment. Issues of communication between NASA and the spacecraft manufacturers were also noted, and there were not sufficient procedures for crew rescue.

Block II

The programme was delayed as changes were made to the CM's design:

- ☐ An outwardly opening hatch was made to open in 7 seconds.
- ☐ Cabin atmosphere at launch changed to 60 per cent oxygen and 40 per cent nitrogen from 100 per cent oxygen.
- ☐ Flammable material was replaced.
- ☐ Better wire insulation was introduced.

It was reckoned by many that the accident allowed the subsequent project's success as it forced a re-evaluation of procedures and the spacecraft's design.

AS-204

The mission was designated AS-204: the fourth flight using the second type of Saturn rocket in the Apollo/Saturn programme, but the crew had patches made saying 'Apollo 1'. After the fire, it was officially given this name on request of the crew's widows.

> We are in a risky business, and we hope that if anything happens to us, it will not delay the program. The conquest of space is worth the risk of life.
>
> Gus Grissom, 1966

○○○○○○○○○○○○○○○○○○○○○○○○○○○○○○○○

Unmanned Tests

Three unmanned test flights were made after Apollo 1.

Mission	Launch Date	Duration	Details
Apollo 4*	9 Nov 1967	8 hours, 37 minutes	First Saturn V flight. A Block I CSM was flown and a dummy LM carried for ballast. The CSM was boosted to an altitude of 11,242 miles (18,092km) then directed back to Earth, emulating a lunar mission re-entry.
Apollo 5	22 Jan 1968	11 hours, 10 minutes	LM first flight, using the Saturn IB assigned to Apollo 1. LM's ascent engine was fired while ascent stage still attached to the descent stage, simulating an in-flight abort.
Apollo 6	4 Apr 1968	9 hours, 57 minutes	Longitudinal 'pogo' vibrations occurred early in the flight and two S-II engines shut down. The S-IVB engine wouldn't restart in orbit.

*Apollo 2 and Apollo 3 were not assigned. Three unmanned Apollo/Saturn flights had been made in 1966 and so it was decided to begin the sequence at 4.

TV

> Our building is shaking. The roar is terrific! The floor is shaking. This big glass window is shaking and we're holding it with our hands!
>
> Walter Cronkite

As Apollo 4 lifted off, the broadcaster could hardly be heard as soundwaves reached the temporary studio 3 miles (4.8km) from the launch pad. Cronkite extensively covered American spaceflight for CBS News.

APOLLO 7

Crew:	Wally Schirra (CDR) Donn Eisele (CMP) Walt Cunningham (LMP)
Mission patch:	Astronaut's surnames surround a CM flying over America
Date (launch):	11 October 1968
Date (return):	22 October 1968
Mission duration:	10 days, 20 hours, 9 minutes
LM:	n/a
CM:	Unnamed
Landing site:	n/a
Time on the Moon:	n/a
Mission objective:	To test the CSM and mission support facilities in Earth orbit

Phoenix
The original patch design included a phoenix. It was vetoed by NASA management who didn't want reminders of the Apollo 1 fire.

1,341

Number of changes approved by the committee overseeing improvements to the CM.

Saturn IB

Apollo 7 used the smaller two-stage Saturn IB rocket. The first stage employed the tanks of eight Redstone rockets and the second stage was a S-IVB. The rocket vibrated so much after lift-off Schirra thought the world was 'coming to an end'. The IB was later used for Skylab and Apollo–Soyuz missions.

Günter Wendt

> I wonder where Günter Wendt?

This pun, created by Donn Eisele, referred to launch pad leader Günter Wendt, one of the last people astronauts saw in the launch tower's White Room before lift-off. Wendt oversaw crew ingress on Mercury and Gemini but as his company, McDonnell, did not manufacture the CM he was not present for Apollo 1. Schirra insisted on him being hired by North American.

Traditionally gifts were exchanged between Wendt and the crews on launch day. He gave Neil Armstrong a key to the Moon and received a card reading, 'Space Taxi. Good between any two planets.'

'Wally's Ship'

Workers at North American nicknamed Apollo 7's CM 'Wally's Ship'. The 45-year-old Schirra had flown on Mercury and Gemini and was a neighbour of Gus Grissom. He took a keen interest in the vehicle that would lift off from the same launch pad where his friend and fellow astronauts had died twenty-one months previously.

'Yabadabadoo!'

Schirra gave this cry, used by cartoon character Fred Flintstone, when the CSM's unexpectedly powerful engine was fired for the first time.

TV

Keep those cards and letters coming in folks.

Card held up during TV broadcast

The crew made the first American TV broadcasts from space. Carrying cameras on board was resisted by some in NASA who thought them unnecessary weight and a distraction for the crew. The seven broadcasts 'From the lovely Apollo room high atop everything' were popularly received. The 'Wally, Walt and Donn Shows' won them a special Emmy award. All subsequent flights included TV broadcasts.

9

The crew took nine large boxes of tissues. None of them had colds when they took off, but Schirra and then his crewmates caught head colds early in the flight. It was more unpleasant than on Earth: nasal passages became blocked and couldn't be cleared properly. Schirra later appeared in television adverts for the decongestant he had used in space.

Problems

I wish you would find out the idiot's name who thought up this test. I want to talk to him personally when I get back down.

Wally Schirra

There were several moments when the crew and Mission Control were at odds. Before launch, Schirra was unhappy as wind conditions were beyond limits. He was concerned if they aborted after lift-off, the wind would blow the parachuting capsule back onto land where the astronauts' couches were not capable of protecting them adequately.

In space, Schirra refused to proceed with a scheduled TV broadcast, and declared he was going to be the 'on-board flight director'. Untested procedures and altered timings from Mission Control were resisted by a commander and crew keen to run a flawless mission.

With the crew still suffering from colds, Schirra refused the order to don helmets during re-entry. He was concerned they could suffer eardrum damage by not being able to equalise the pressure. The helmets were kept off.

Schirra had served in the US Navy where a commander had ultimate authority over his craft. Many in Mission Control thought differently. Schirra was described by Gene Kranz as the 'grumpy commander'. Senior manager Chris Kraft vowed none of the three would fly again. Schirra had indicated his intention to resign before the flight; Eisele and Cunningham remained on the flight roster but never flew in space again.

163

The crew made 163 orbits. The flight's schedule of activities was frontloaded so important tests were performed first in case the mission was ended early. By the end of the mission Eisele described his crew members as suffering from 'cabin fever'.

56

Out of the fifty-eight tests and experiments all but two were fully or partially achieved. The spacecraft's electrical, environmental and guidance systems were tested and rendezvous with the booster was practised.

ΛPOLLO 8

Crew:	Frank Borman (CDR)
	Bill Anders (LMP)
	Jim Lovell (CMP)
Mission patch:	Astronauts' names on a numeral '8' which circles the Earth and Moon
Date (launch):	21 December 1968
Date (return):	27 December 1968
Mission duration:	6 days, 3 hours, 1 minute
LM:	n/a
CM:	Unnamed
Landing site:	n/a
Time on the Moon:	n/a
Mission objective:	To fly to the Moon, orbit, then return to Earth

Original Plan

Are you out of your mind?

James Webb, NASA Administrator, on being told of the proposal

Apollo 8 was originally to be a test of the LM and CSM in Earth orbit but delays in LM production meant this wasn't possible. George Low, manager of the Apollo Spacecraft Program Office, suggested that instead the mission could be a flight to the Moon, using just the CSM. There were concerns the USSR might attempt a Moon mission with a manned Zond craft in late 1968. The bold plan was accepted. The flight was offered to Jim McDivitt but he turned it down and Frank Borman, whose crew were next in line to fly, accepted.

Pioneers

The day before launch the crew had lunch with Charles Lindbergh, whose solo flight from the USA to Paris had caught the imagination of the world four decades previously.

TLI

Apollo 8 – you're Go for TLI.

With these undramatic words Capcom Mike Collins gave clearance for Apollo 8 to begin its journey towards the Moon. TLI was 'Trans-Lunar Injection' – the burning of the S-IVB's engine to accelerate the spacecraft to 24,226mph (38,988km/h). The speed was not 'escape velocity' – i.e. that required to completely leave Earth's gravitational field – but was enough to reach the Moon's. If the Moon was missed the spacecraft would continue on a large, elliptical orbit before returning to Earth.

Collins had been a member of the original Apollo 8 crew before a medical problem saw him replaced by Jim Lovell.

66 Hours

After TLI the spacecraft coasted towards the Moon for 66 hours. The CSM started slowing down as soon as its engine cut off and by the time it reached the Moon's 'sphere of influence' – i.e. where the Moon's gravitational field was dominant as opposed to the Sun or Earth's: 215,000 miles (346,000km) from Earth – it had slowed to 2,000mph (3,200km/h). The Moon's gravitational pull then accelerated the spacecraft before it entered orbit.

853 Miles

The furthest distance from Earth a manned spacecraft had reached before Apollo 8 was 853 miles (1,372km), with Gemini 11 in 1966. Apollo 8 travelled over 280 times further.

Space Sickness

Borman was sick early on in the flight and also suffered from diarrhoea. While not only unpleasant for him and his crewmates in the spacecraft's small confines, it also presented concerns for the progression of the flight. Borman kept his illness quiet for a while, determined to press on with the mission.

Loss of Signal

A major event in the flight was when the spacecraft went behind the Moon – the first time humans were out of contact from Earth, as radio signals couldn't reach them. The guidance controllers' predictions were so accurate – to the second – that Borman jokingly suggested Mission Control had turned the transmitter off.

Impressions

As the astronauts orbited 69 miles (128km) above the surface, they gave their immediate impressions:

Borman: 'It is a vast, lonely, forbidding-type existence, or expanse of nothing … it certainly would not appear to be a very inviting place to live or work.'

Lovell: 'The vast loneliness up here of the Moon is awe inspiring.'

Anders: 'A very rather dark and unappetizing looking place.'

Earthrise

It seemed ironic to me that we really came to discover the Moon and yet when we saw the Earth rise we really discovered the Earth.

Bill Anders, speaking in 2009

On their fourth orbit the crew saw something spectacular: the Earth rising above the lunar horizon. They quickly took photographs and one by Anders became an iconic image of the twentieth century. Prints were sent to world leaders, including Ho Chi Minh, President of North Vietnam. Despite his country being at war with America, Ho Chi Minh sent a card to US President Lyndon Johnson with a thank-you message.

Earthrise.

Genesis
The crew gave a TV broadcast from lunar orbit on Christmas Eve and it was felt something appropriate should be said. After describing the newly discovered world passing beneath them, Anders, Lovell and Borman took turns to read from the Book of Genesis. An audience estimated at between 0.5 and 1 billion people heard Borman end the reading:

> And from the crew of Apollo 8 we close with good night, good luck, a Merry Christmas and God bless all of you, all of you on the good Earth.

Xmas Present

Please be informed there is a Santa Claus.

Jim Lovell

A critical moment of the mission was the engine burn to take the spacecraft out of lunar orbit and send the crew homewards. Lovell's topical statement confirmed the CSM's engine had fired correctly, accelerating them in just under 3½ minutes from 3,643mph (5,862km/h) to 6,029mph (9,702km/h). Apollo 8 spent 20 hours and 7 minutes around the Moon.

On their way home, Jack Schmitt read out a parody of the seasonal poem *T'was the Night Before Christmas*, which began:

T'was the night before Christmas and way out in space,
The Apollo 8 crew had just won the Moon race.

Driving

As they coasted towards Earth, Capcom Mike Collins passed on a question from his son: who was driving? Anders responded, 'I think Isaac Newton is doing most of the driving right now.'

Christmas in Space

While those on Earth celebrated Christmas, those in space did not wholly miss out. Turkey with stuffing and cranberry sauce had been provided, along with small bottles of brandy – deemed off limits by Borman. There were also gifts from the crew's wives, with cufflinks and tie pins being received. Lovell had arranged for a Christmas gift to be delivered to his wife with a note from 'The man in the Moon'.

Men of the Year

For their historic mission the crew were named Men of the Year by *Time* magazine.

Telegram

Thank you, Apollo 8. You saved 1968.

An anonymous telegram was received by the crew after the flight. The year had seen violent protests, political assassinations and the escalating war in Vietnam. Apollo 8's voyage at the end of the year was seen by many to offer hope for the future.

Verne and Apollo

Jules Verne's 1865 story *From the Earth to the Moon* describes intrepid travellers journeying in a spacecraft fired into space by a huge cannon. While this method of propulsion was not replicated in Apollo a century later, there are a number of similarities that led Frank Borman to write: 'Who can say how many of the world's space scientists were inspired, consciously or unconsciously, by their boyhood reading of the works of Jules Verne?'

Some of the similarities:

❑   Lunar flights launch from Florida.
❑   Millions of spectators flock to see the launches.
❑   Three men travel in the capsules.
❑   The returning vehicles land in the Pacific Ocean.
❑   There is a character in Verne's book called Armstrong.
❑   The cannon that fired Verne's characters was called *Columbiad*. Apollo 11's CM was *Columbia*.

2001: A Space Odyssey

This epic science-fiction film was released in April 1968 and many involved in the space programme went to see it. In one scene a black monolith, placed millions of years ago by extra-terrestrials, is discovered on the Moon. Apollo 8's astronauts later told author Arthur C. Clarke that when they saw the Moon's far side for the first time, they'd toyed with the idea of announcing the sighting of a black monolith. On Apollo 13, the announcement of the explosion – on the CM *Odyssey* – mirrored a line in the film's script when computer HAL utters, 'I think we've got a problem.'

APOLLO 9

Crew:	Jim McDivitt (CDR) Rusty Schweickart (LMP) Dave Scott (CMP)
Mission patch:	Inside the astronauts' names and mission number, a LM and CM circle a Saturn V
Date (launch):	3 March 1969
Date (return):	13 March 1969
Mission duration:	10 days, 1 hour, 1 minute
LM:	*Spider* (its spider-like shape)
CM:	*Gumdrop* (when delivered it was covered in a blue material, resembling a sweet wrapper)
Landing site:	n/a
Time on the Moon:	n/a
Mission objective:	Test of spacecraft, systems, and docking and rendezvous procedures in Earth orbit

Test Mission
The flight was seen as a test pilot's dream as it included many items being evaluated in space for the first time, such as the spacesuits, the LM, the extraction of the LM from its adapter by the CSM and rendezvous between the spacecraft. It had to be without major failure or the end of the decade target for the first lunar landing would be missed.

Delayed
When the crew caught head colds, the launch was delayed by three days.

Staging
The accelerative forces had caused the Saturn to become slightly compressed in length. When the first stage engines stopped, it attempted to spring back into shape and this, combined with the deceleration from 4g to 1g, threw the astronauts forward. Scott and Schweickart almost hit the control panel with their helmets.

Sickness
Schweickart was to demonstrate the Apollo spacesuit but suffered a bout of sickness. As being ill while wearing the spacesuit was potentially fatal, the EVA was cancelled. When he recovered it was reinstated but curtailed: instead of demonstrating an emergency transfer between the LM and CM, Schweickart stood on the LM's porch. He later spent three months being evaluated to discover the reasons behind what became known as space adaptation syndrome.

Schweickart's Experience
During the EVA, Schweickart was told to halt activity while attempts were made to fix a jammed camera. This allowed him 5 minutes to gaze at the Earth. Years later he expressed how powerful an impact this had, leading him to try to answer questions about existence.

Separation
The LM and CSM undocked to perform rendezvous manoeuvres and separated to a distance of over 100 miles (161km) before converging. The LM was put into a different orbit to test its descent engine, then staging procedures – the LM's ascent stage separating from the descent stage – were carried out.

Success

The mission was a success and meant the next flight would be to the Moon.

APOLLO 10

Crew:	Tom Stafford (CDR)
	Gene Cernan (LMP)
	John Young (CMP)
Mission patch:	'Apollo' and the astronauts' surnames surround the Moon, Earth, a LM and a CSM flying through a Roman numeral X
Date (launch):	18 May 1969
Date (return):	26 May 1969
Mission duration:	8 days, 0 hours, 3 minutes
LM:	*Snoopy* (after the *Peanuts* cartoon character and also because it was to 'snoop' around the Moon)
CM:	*Charlie Brown* (the *Peanuts* cartoon character who looked after Snoopy)
Landing site:	n/a
Time on the Moon:	n/a
Mission objective:	Final lunar landing rehearsal

Moon Landing?

It was thought by some, including George Mueller, head of NASA's Office of Manned Space Flight, that Apollo 10 should make the first attempt at a landing. However, the LM assigned was too heavy and there were concerns about not knowing enough about gravitational conditions in lunar orbit. The LM had only been flown once before and only one mission had flown to the Moon.

Staging

That staging was quite a sequence!

Stafford's comment came after the jettison of the first stage, which saw the crew rapidly pushed forward and backwards four times. The astronauts were concerned the vibration had caused the spacecraft structural damage. The engine burn for TLI also caused high-frequency vibrations that caused the crew to think the mission would be curtailed.

Flat Earth

You can tell the members of the British Flat Earth Society they are wrong: the Earth is round.

Tom Stafford

The president of the society responded saying while it may be round it was still a disk.

47,000ft

We is down among them, Charlie!

Gene Cernan to Capcom Charlie Duke

Once in lunar orbit *Snoopy* undocked and descended to a height of 47,400ft (14,448m) above the surface, although they passed closer to the lunar mountains. Cernan and Stafford were the first to see the Moon from such low altitudes and at such high speed: they were travelling at 3,700mph (5,955km/h).

LM Staging

Son of a bitch! What the hell happened?

Gene Cernan

The LM passed over Apollo 11's proposed landing site at the lowest point of its orbit. It then rose before descending again, to allow a simulation of a lunar surface launch. As Stafford and Cernan prepared to discard the descent stage, their spacecraft went out of control for several seconds before control was regained. Cernan's outburst went live to the world and he was criticised for profanity. The cause of the incident was put down to switches being set incorrectly.

Moon Music

As the LM flew towards the CSM around the far side of the Moon, to dock and return home, the crew in both spacecraft reported hearing 'weird music' which lasted for about an hour. The strange whistling noise, described by Cernan as an 'outer space-type thing', was thought to be interference from the spacecrafts' radios.

24,791mph

Apollo 10's CM was the fastest of all Apollo spacecraft, arriving for re-entry at 24,791mph (39,897km/h). The CM was travelling at over 36,000ft per second (10,900mps), equivalent to reaching an (Earth) airliner's typical cruising altitude from the ground in 1 second.

APOLLO 11

Crew:	Neil Armstrong (CDR) Buzz Aldrin (LMP) Mike Collins (CMP)
Mission patch:	'Apollo 11' above the Earth and an eagle carrying an olive branch, descending to the lunar surface
Date (launch):	16 July 1969
Date (return):	24 July 1969
Mission duration:	8 days, 3 hours, 18 minutes
LM:	*Eagle* (from the patch design)
CM:	*Columbia* (reflecting Columbus's voyage of 1492, Jules Verne's Moon voyage book and a symbol of America)
Landing site:	Mare Tranquillitatis (Sea of Tranquility)
Time on the Moon:	21 hours, 36 minutes
Mission objective:	First attempt at lunar landing

The Crew

> Tis likely enough that there may be means invented of journeying to the Moon, and how happy they shall be, that are first successful in this attempt?
> John Wilkins, 1640

Mike Collins described his fellow crew members as 'amiable strangers'. While they never gelled like some crews, they still carried out their tasks effectively. Armstrong and Aldrin were both serious-minded astronauts, while Collins's humour and light heartedness masked the former test pilot's capabilities.

LΛUNCH

Launch Day Timeline

Time (a.m.) US Eastern Daylight Time	Details
4.15	Crew woken, then medical check-up.
5.00	Breakfast (steak, scrambled eggs, orange juice, toast).
5.35	Suiting up.
6.27	Transport to launch pad commences, using Astronaut Transfer Van.
6.54	Ingress to CM begins.
7.25	CM hatch is closed.
9.27	Access arm retracts.
9.32	Lift-off.

417
Back-up crew member Fred Haise ran through 417 items on the spacecraft's checklist before the prime crew arrived. He remained inside the CM to help the crew ingress.

Million
An estimated 1 million people came to Florida to watch the launch. At times drivers used the empty opposing lanes to get near to the Cape.

Drinks
As crowds gathered in Cocoa Beach to watch the launch two cocktails were popular: the Liftoff Martini and the Moonlander, which consisted of vodka, crème de menthe, crème de cacao, soda and lime.

3,493

The mission generated huge amounts of interest from around the world; there were 3,493 reporters at KSC for the launch.

3 1/2 miles

KSC's viewing site was positioned 3½ miles (5.6km) from the launch pad as this was calculated, in the event of a catastrophic event, to be far enough away from any falling debris weighing over 100lb (45kg).

> This is the last day of the old world.
>> Science fiction writer Arthur C. Clarke, who witnessed the launch

Countdown

> 30 seconds and counting. Astronauts report 'It feels good.' T minus 25 seconds. Twenty seconds and counting. T minus 15 seconds, guidance is internal. Twelve, eleven, ten, nine, ignition sequence starts … six, five, four, three, two, one, zero, all engines running. Lift-off! We have a lift-off, 32 minutes past the hour. Lift-off on Apollo 11.

Jack King, NASA Public Affairs Officer, gave the final moments of the countdown. King was known as the 'Voice of Apollo'.

JOURNEY TO THE MOON

> That Saturn gave us a magnificent ride.

Armstrong after the spacecraft sent them towards the Moon, 2 hours, 53 minutes into the flight.

Moon Rabbit

On their way to the Moon, the crew were told to look out for a 'lovely girl with a big rabbit' – Chang'e, the Chinese Moon goddess and her companion, who are believed to be seen crushing up medicine on the lunar surface. Mike Collins responded they would keep a close eye out for the 'bunny girl'.

Luna 15

Three days before Apollo 11 launched, the USSR began their own bid for lunar success. Concerns that Luna 15's trajectory could affect the American spacecraft were allayed when the Soviets gave NASA details of its flightpath. Luna 15 didn't begin its descent until after Armstrong and Aldrin had completed their EVA and contact with the lander was subsequently lost.

'Czar of the Ship'

USSR state newspaper *Pravda* gave Apollo's commander this title.

UFO

On the third day out, Armstrong asked Mission Control the location of the S-IVB stage. He was told it was 6,900 miles (11,104km) away. The crew had seen an object and thought it was the stage but this meant it was too far away. It was variously described as two joined rings, a hollow cylinder or an 'open book' L-shaped object. Mike Collins said after the flight, 'It was really weird.' Buzz Aldrin later thought it was one of the panels that had housed the LM.

Flashes

Another unusual event was the observing of flashes inside the spacecraft. Both Armstrong and Aldrin reported seeing them with their eyes open or closed. It was thought to be charged particles impacting on their retinas. Apollo 11 was the first mission to have seen these but later missions also reported them.

OOOOOOOOOOOOOOOOOOOOOOOOOOOOOO

THE LΛNDING

What to Expect?
There was still some anxiety over what would be encountered. In 1955 astrophysicist Dr Thomas Gold of Cornell University predicted the Moon was covered in a layer of thick dust (given the name 'Gold's Dust') and anything landing would immediately sink. The robotic landers of the early 1960s alleviated most of these fears. It was also thought by some that hydrogen-rich lunar rocks would ignite on contact with the oxygen in the LM. Another fear that wasn't given quite as much credence was from a man in Israel who warned NASA not to land as giant ants lived on the Moon.

50:50
Both Armstrong and Collins thought they had a 50:50 chance of a successful mission. If they were forced to abort, NASA management had promised the crew a second opportunity to attempt the landing.

The Descent
Before the descent, Apollo 11's two spacecraft were in an orbit measuring 75.6 miles (121.6km) at its highest point and 61.9 miles (99.6km) at the lowest. After separation, *Eagle* carried out a 28.5-second-long Descent Orbit Insertion (DOI) burn of its engine to take it down to an altitude of 50,000ft (15,240m).

It then started Powered Descent Initiation (PDI) with *Eagle*'s engine slowing the craft from its orbital speed of 3,700mph (5,955km/h). The astronauts could see the surface at the start to ascertain their position, then lost sight of it as the spacecraft rotated to allow the landing radar to measure altitude and speed. This meant the crew were looking upwards, away from the Moon. Later, *Eagle* pitched forwards to allow the crew to see the surface, and their landing site.

Landing ∧pproach
The landing site had been overflown by Apollo 10 and by Apollo 11 on their initial orbits. There were several features that served as landmarks to indicate they were on the correct approach:

- ❑ Maskelyne Crater
- ❑ Diamondback Rille
- ❑ Sidewinder Rille
- ❑ Boot Hill
- ❑ The Gashes
- ❑ Last Ridge
- ❑ Duke Island
- ❑ Mount Marilyn
- ❑ US.1 (Hypatia Rille)

Issues
Eagle's descent to the surface was not straightforward and the crew faced several issues that had the potential to cause the landing to be aborted.

Communications
Communications between *Eagle* and Mission Control were subject to disruption, and the astronauts yawed the spacecraft and also changed the antennae being used to ensure they remained in contact during this critical phase.

Landing Site
Armstrong noticed the LM was reaching landmarks ahead by 3 seconds, which meant it would miss its intended landing site. It was thought remaining air pressure in the tunnel between the LM and CM at undocking had caused the issue.

Computer Alarms

During the powered descent, several computer program alarms were sounded. These were designated '1202' and '1201' and indicated the on-board computer was being overloaded. As the crew waited anxiously, having no knowledge of what these meant, Mission Control quickly stated the flight should continue. The computer had been designed to continue to work in such situations, with each task speedily evaluated by priority scheduling.

Boulders

As *Eagle* neared the surface, Armstrong saw the computer was bringing them to land in a boulder field next to a large crater. He took control while 500ft (152m) above the surface and looked for an alternative spot. Mission Control saw he was flying horizontally but Armstrong was fully concentrating and had no opportunity to inform them of his actions. He then had to fly over another crater before finally landing the LM.

Fuel

There were concerns that with the extra manoeuvring they were running low on fuel. Capcom Charlie Duke called '60 seconds' while *Eagle* was below 75ft (23m) above the lunar surface, meaning they had 1 minute of fuel remaining before an abort had to be called. However, they were in 'Dead Man's Zone' – there wouldn't be enough time to perform an abort at this height. It was the critical part of the whole Apollo programme so far. Although externally silent, Armstrong was confident of landing by this point and thought the LM could drop successfully from around 40ft (12m) even if they ran out of fuel. It was later thought *Eagle* landed with around 50 seconds of fuel remaining.

Dust

As they neared the surface, dust was blown up by *Eagle*'s engine. This obscured Armstrong's view of the exact spot to land and also in gauging the LM's sideways movement. It was important to land vertically so as not to damage the landing legs.

'Contact Light'

Sensing probes, measuring 68in (1.73m) long, were positioned underneath the landing legs. When they touched the surface they triggered a blue light on the control panel. Aldrin, who had called out descent rates and altitudes throughout the final phases of landing, now indicated the historic landing was only moments away.

Landing

> Houston, Tranquility Base here.
> The *Eagle* has landed.

Armstrong's words announced to the world the landing's success. Duke responded, reflecting the relief of all those in Mission Control – and in the wider outside world – by saying, 'Roger, Tranquility. We copy you on the ground. You've got a bunch of guys about to turn blue. We're breathing again. Thanks a lot.'

The Haystack has Landed?

As close as a month before launch the spacecraft had different names: *Snowcone* for the CM and *Haystack* for the LM, before *Columbia* and *Eagle* were chosen. Other names considered by the crew included:

- ❑ *Antony* and *Cleopatra*
- ❑ *Romeo* and *Juliet*
- ❑ *Chloe* and *Daphnis*
- ❑ *Amos* and *Andy*
- ❑ *Castor* and *Pollux*
- ❑ *David* and *Goliath*
- ❑ *Owl* and *Pussycat*
- ❑ *Majestic* and *Moon Dancer*

156bpm

Armstrong's heart rate during the final stages of the descent was 156 beats per minute.

12 Minutes, 34 Seconds

Eagle took just over 12½ minutes from the start of powered descent, at 50,000ft (15,240m) altitude and 250 miles (402km) distance, to landing. This was 36 seconds longer than planned due to Armstrong's decision to avoid landing in the boulder field.

Landing Site

Consideration had gone into selecting areas that had smooth and level surfaces and presented no obstacles to an easy approach path. Information had been gathered from the Ranger, Surveyor and Lunar Orbiter missions. *Eagle* landed 4 miles (6.4km) further on from and almost a mile (1.6km) to the left of its intended spot.

Immediate Plans

Armstrong and Aldrin had little time to relax and take in the view. They immediately began simulating lift-off in case they had to depart early. With the go-ahead to stay they began preparing for the historic EVA.

A 4-hour rest period had been scheduled but this was cancelled. The astronauts felt unable to rest properly with such a momentous event ahead.

EVA

First Man

I figured the commander ought to be the first guy out.

Deke Slayton

There was much attention on who would be first to walk on the Moon. Initial plans for the EVA indicated the LMP would be first out, as in Gemini where the spacecraft commander remained inside. Bill Anders later revealed he and Alan Bean, who as LMPs were given the job of preparing LM checklists, cheekily wrote 'LMP Egress' before 'CDR Egress'.

For Apollo 11 this would mean Buzz Aldrin would be first. However, management felt Armstrong would be a better candidate for being the figure to make history. The reason given was the LM hatch door opened towards where Aldrin stood and there was no room for him to get around the door and egress first while wearing a bulky spacesuit and backpack. However, Alan Bean later said that this could be gotten around easily by the astronauts swapping places before suiting up.

Aldrin, who was more conversant with the experiments to be carried out on the lunar surface, and had thought he would be first, was disappointed and lobbied to change the procedure. He was unsuccessful and so Armstrong would be first.

First Words

Once the LM's cabin had been depressurised the hatch was opened and Armstrong carefully edged his way out before descending the ladder. At the bottom, standing on the footpad, he paused, then announced he was 'about to step off the lem'. Holding onto the ladder with one hand, he touched the lunar surface with his left foot. At 2.56 a.m. GMT he uttered one of the most famous lines in human history:

> That's one small step for [a] man,
> one giant leap for mankind.

There was much discussion over whether he had inadvertently missed out the 'a' from the first part of the sentence and Armstrong himself wasn't sure. There had been great interest in what he would say at this historical moment. Armstrong said he only thought of the words while on the Moon; however, in a 2013 BBC documentary, his brother Dean revealed he had composed his famous sentence while still on Earth.

Aldrin Egress
20 minutes after Armstrong, Aldrin made his own way out. Before descending the ladder he joked about making sure not to lock them out of the cabin. Once on the surface he described what he saw in an evocative phrase: 'Magnificent desolation'.

Aldrin's Firsts
Aldrin achieved his own 'first' when shortly after landing he took Communion, with a small amount of wine and a wafer. He did not announce to the world what he was doing but did ask for everyone listening to pause and give thanks. Soon after emerging from the LM, Aldrin carried out another first: the first urination on the Moon.

TV

> We've got a picture on the TV.
>
> Bruce McCandless, Capcom

Armstrong's first step was captured by a black and white TV camera, mounted on the LM. Footage from the EVA had a ghostly appearance because of the camera's technical limitations in recording bright images.

13
Armstrong later said that out of ten, he rated the descent to landing on the Moon as thirteen. Carrying out the EVA was a one.

Footprint.

Surface Experiments

One of the first things Armstrong did was collect the contingency sample of lunar soil and rocks in case they had to leave the Moon quickly. A core sample was also taken.

The Early Apollo Scientific Experiment Package (EASEP) was a cut-down version of the later Apollo Lunar Surface Experiments Package (ALSEP). Four experiments were deployed:

Passive Seismic Experiment

Detected seismic activity and meteoroid impacts. The seismometers sent data for three weeks after the astronauts' departure. The seismometer was so sensitive it recorded the astronauts' backpacks being discarded before launch.

Laser Ranging Retro-Reflector Experiment
This array of 100 fused silica cubes reflected laser beams sent from Earth and was used to accurately measure distances to and from the Moon.

Solar Wind Composition Experiment
Made of aluminium foil, it collected atomic particles emitted from the Sun and was returned to Earth for analysis.

Lunar Dust Detector Experiment
Measured the reduction in solar cell power caused by dust accumulation.

Unplanned Activity
While the astronauts' time was heavily restricted by the timeline, there was one moment of unplanned activity. Armstrong ran 197ft (60m) towards Little West crater, which they had overflown on the landing approach. He considered entering the crater to retrieve what would be valuable rock specimens but erred on the side of caution.

The astronauts were instructed not to move too far away from the LM – the whole area of their EVA could fit inside a football stadium.

Magazine 40/S
Many of the mission's iconic images were taken on one 70mm Kodak film magazine, attached to a Hasselblad camera. It contains all photographs taken on the lunar surface.

Armstrong took most of the photographs and amongst the 122 images taken there are no clear images of the 'first man' on the lunar surface, save for several silhouettes he took of himself and a photograph of him in shadow near the LM. Aldrin was about to photograph his companion saluting the flag when President Nixon telephoned.

Man on the Moon.

2 hours, 31 minutes

> There wasn't time to savour the moment.
>
> Buzz Aldrin

It was only 2 hours 31 minutes from the hatch being opened to it being closed. Apollo 11 was the only mission to have one EVA.

Moon Dust
When the astronauts removed their helmets back in the LM, Armstrong thought the Moon dust smelled of wet ashes and his companion described it as like gunpowder.

Items Carried to the Moon
Various items were taken to the Moon, some official and others as part of the astronaut's own personal belongings, kept in Personal Preference Kits (PPK). Many of the 'flown' objects were subsequently given as gifts to family, friends and colleagues.

Official
Medals honouring Yuri Gagarin and Vladimir Komarov, two cosmonauts killed in flying and space accidents.*
US flags (to be presented to US Senate and Congress).
Flags of all fifty American states.
Apollo 1 mission patch.*

Armstrong
Flags (including Purdue University and US flag).
Medallions.
College fraternity pin.
Pennant for Argentinean football team Independiente de Avellaneda.
World Scout badge.

Diamond astronaut pin given to Deke Slayton by widows of Apollo 1 crew.
Piece of fabric and propeller from Wright brothers' 1903 Flyer.

Aldrin
Commemorative first day covers.
Flags.
Medallions.
Rings.
Gold bracelet that had belonged to his mother.
Four olive branch pins: one for each crew member and one to be left on lunar surface.
YWCA pin.

Collins
Flags (Including USA, US Air Force and District of Columbia).
Medallions.
Tie pins.
Cufflinks.
Gold crosses.
Gold locket.
Coins.
Brooches.
College ring.
Nappy ring.
A hollow bean containing fifty tiny carved ivory elephants.

* Left on lunar surface.

Ceremonial
Several symbolic acts were carried out.

Plaque

A steel plaque was affixed to one of the LM's landing legs. It was inscribed with representations of the Earth and signatures of the crew and President Nixon. The inscription was read out:

> Here men from the planet Earth
> first set foot upon the Moon,
> July 1969 AD.
> We came in peace for all mankind.

Flag

NASA decided an American flag was to be placed at the landing site but it could not be seen as a symbol of American claims on the territory as the United Nations' Outer Space Treaty prevented any nation doing so.

The flag was given an aluminium crossbar as with no wind 'Old Glory' would just hang limply. As it was a late addition, there was little time to practise and the astronauts experienced difficulty in driving the flagpole into the hard ground underneath the loose surface soil. Their anxieties that it might embarrassingly fall over were not realised.

Disk

A small silicon disk contained messages of support from seventy-three world leaders, US Presidents and the names of various American politicians and NASA officials. The messages were reduced in size by a factor of 200 in order to fit on the disk, using techniques developed in the computing industry.

Many messages included wishes that the world would benefit from the endeavour and that it would help bring peace. One message stated:

> The Government and people of Trinidad and Tobago acclaim this historic triumph of science and the human will. It is our earnest hope for mankind that while we gain the Moon, we shall not lose the world.
>
> Eric Williams, Prime Minister

Phone Call

> This certainly has to be the most historic telephone call ever made from the White House.
>
> President Richard Nixon

The EVA was interrupted by an unexpected call from President Nixon. Nixon later flew to the aircraft carrier USS *Hornet* to greet the returning astronauts.

LUNAR ORBIT

What did Mike do?
When his crewmates descended to the surface, Mike Collins was left alone in lunar orbit. He was unable to hear his crew mates' 'first steps' as the CM had gone behind the far side, cutting out communications. Collins enjoyed his 22 hours of solitude, to which aviation pioneer Charles Lindbergh described in a letter to him as 'an aloneness unknown to man before'.

Tasks carried out by Collins in orbit:

- ❏ Navigation readings.
- ❏ Equipment and systems checks.
- ❏ Relaying communications between *Eagle* and Earth.
- ❏ Searching for *Eagle*. As it had not landed where intended, there was much effort into locating the LM. Collins used the sextant to peer down but was unsuccessful.

18

Collins had practised eighteen different procedures to rendezvous with *Eagle* if lift-off did not go to plan. It was his biggest concern of the whole flight. If it failed, he would return to Earth on his own and become, as he described it, 'a marked man'.

RETURN TO EΛRTH

Λscent

While on the Moon the crew found a broken-off circuit breaker switch, caused by their large backpacks manoeuvring in the small cabin. The switch was needed to ignite the ascent engine and so Aldrin used a felt-tip pen. He kept both as souvenirs.

In the Event ...

Had Armstrong and Aldrin become stranded on the Moon, a speech had been prepared for President Nixon. It began with this paragraph:

> Fate has ordained that the men who went to the Moon to explore in peace will stay on the Moon to rest in peace. These brave men, Neil Armstrong and Edwin Aldrin, know that there is no hope for their recovery. But they also know that there is hope for mankind in their sacrifice. These two men are laying down their lives in mankind's most noble goal: the search for truth and understanding.

Quarantine

The lift-off and rendezvous were successful and the CM returned to Earth without incident. The crew were put into a 3-week quarantine, in case they had returned with any extra-terrestrial organisms that could harm human life.

Eagle's ascent stage approaches *Columbia*.

The astronauts donned Biological Isolation Garments which were thrown into the CM's hatch by the recovery crew after splashdown. The crew were then transferred to a Mobile Quarantine Facility on board the aircraft carrier. This was flown to Houston and the astronauts transferred to the Lunar Receiving Laboratory for the rest of the quarantine period. Quarantine was discontinued after Apollo 14.

Customs Declaration

On their return the crew had to submit a General Declaration form to the US customs authority. The form included these pieces of information:

Flight No.:	Apollo 11
Date:	24 July 1969
Departure:	Moon
Arrival:	Honolulu, Hawaii
Cargo:	Moon rock and Moon dust samples
Any other condition on board which may lead to the spread of disease:	To be determined

Expenses Claim
After the flight Aldrin submitted a travel expenses claim for the dates 7 to 27 July. It included the use of 'Government spacecraft' from Houston to Cape Kennedy, Moon, Pacific Ocean, Hawaii and return to Houston. He received $33.31.

THE WΛTCHING WORLD

600 Million
An estimated fifth of the world's population watched the first-ever Moonwalk on TV. Ninety-four per cent of US televisions were tuned to stations broadcasting the event.

32 Hours
US television network CBS broadcast continuously for 32 hours. Walter Cronkite presented twenty-seven of these. One of the guests was science-fiction author Robert Heinlein, who said:

> This is the greatest event in all of the history of the human race up until this time.

Headlines

Man takes first steps on the Moon

The Times

Moon walk!

Herald Examiner (Los Angeles)

Neil Steps On The Moon

Wapakoneta Daily News

Men Walk On Moon
Astronauts land on Plain: collect rocks, plant flag

The New York Times

Man On The Moon

Daily Mirror

Man Walks On Moon

The Vancouver Sun (renamed *The Moon!* for the occasion)

They did it!
World holds breath for a safe return
Two Moon-walkers stride like giants

The Province (Vancouver)

First Footprints On An Alien World
Mark A 'Giant Leap for Mankind'
Moon is a 'Magnificent Desolation'

Omaha World-Herald

Yanks Walk on Moon!
Heroes Safe. Whole World Cheers

Rocky Mountain News

Deux Terriens Ont Marche Sur La Lune

Paris Jour

Fantastico!

El Heraldo De México

AFTERMATH

I just want to remind you that the most difficult part of your mission is going to be after recovery.

Jim Lovell to the crew before re-entry

4 Million
Four million New Yorkers greeted the crew in a ticker-tape parade on 13 August 1969. On the same day they took part in a parade in Chicago before attending a celebratory dinner hosted by President Nixon along with 1,400 other guests in Los Angeles.

100 Million
100 million people saw the astronauts and their wives on the 'Giant Leap' goodwill tour of twenty-three countries in 37 days in the autumn of 1969. Neil Armstrong made another tour later in the year, accompanying comedian Bob Hope to visit American service personnel in Vietnam and other countries.

Neil, we missed the whole thing.

Aldrin to Armstrong, on catching up on the coverage of the mission once back on Earth

In Honour
Neil Armstrong was honoured in several ways. They include:

❑ Neil Armstrong Operations and Checkout Building, Kennedy Space Center, Florida, USA
❑ Neil Armstrong Hall of Engineering, Purdue University, Indiana, USA
❑ Neil A Armstrong Flight Research Center, Edwards Air Force Base, California, USA
❑ Auglaize County Neil Armstrong Airport, Ohio, USA
❑ Neil Armstrong Elementary School, Bettendorf, Indiana, USA
❑ Neil Armstrong Middle School, Bethel Park, Pennsylvania, USA
❑ Armstrong Park, Carol Stream, Illinois, USA
❑ The Neil Armstrong Way, Tralee, Ireland
❑ Neil Armstrong Avenue, Herndon, Virginia, USA
❑ R/V *Neil Armstrong*, research vessel.

Space Window
On 21 July 1974 a Space Window at the Washington National Cathedral was dedicated, with the crew in attendance. A piece of Moon rock weighing 0.25oz (7.18g) was included in the stained glass.

Eisenhower Dollar
In 1971 the US Mint released a $1 coin. It featured former President Dwight D. Eisenhower on one side and the reverse had Apollo 11's eagle and the Moon.

Apollo High School
This school in Glendale, Arizona, was founded in 1970. Its logo features an eagle similar to Apollo 11's. Astronaut Jack Swigert was present at the school's dedication ceremony. A recent teacher was called Mark Luna.

Module McGhee
A girl born on 23 July 1969 was named Module in honour of the mission. Her mother claimed it was her husband's idea and she had to veto calling their new arrival 'Lunar Module McGhee'. Other children born at the time were given the name Apollo.

Moon Cheese
Many commercial companies made the most of the opportunity by marketing products linked to the event. A cheese manufacturer in Neil Armstrong's home town produced 'Moon Cheeze', sold in packaging featuring an astronaut.

Apollo Surface Experiments
Scientific instruments were deployed on the surface in each mission. Most formed part of the Apollo Lunar Surface Experiments Package (ALSEP) first used on Apollo 12, following Apollo 11's EASEP.

ALSEP experiments were connected via cables to a central station which transmitted data back to Earth. It was powered by a plutonium-powered thermoelectric generator. The power required was equivalent to that for a 75-watt light bulb.

The last experiment was switched off in 1977, having run for longer than envisaged.

Name	Acronym	Details	Missions
Active Seismic Experiment	ASE	A thumper device was used to fire nineteen charges into the ground. The resulting seismic waves were measured by geophones to provide insights into the Moon's structure. Grenades were also fired by mortar, up to 2,950ft (900m) in distance, once the astronauts had left.	14, 16
Cold Cathode Ion Gauge	CCIG	Detected the lunar atmosphere. As there was so little it was easily affected by external factors. The gauge detected outgassing from the astronaut's spacesuits when they came near.	12, 14, 15
Charged Particle Lunar Environment Experiment	CPLEE	Measured solar particles close to the lunar surface.	14
Cosmic Ray Detector Experiment	CRD	Deployed on the LM then retrieved at the end of surface activities and returned to Earth.	16, 17
Heat Flow Experiment	HFE	Probes measured the amount of heat lost from interior of the Moon.	15, 16, 17
Lunar Atmosphere Composition Experiment	LACE	A mass spectrometer detected the composition of the Moon's atmosphere.	17
Lunar Dust Detector	LDD	Photocells measured dust that gathered on the EASEP/ ALSEP's Central Station.	11, 12, 14, 15
Lunar Ejecta and Meteorites	LEAM	Measured micrometeorites and lunar ejecta particles.	17
Lunar Portable Magnetometer Experiment	LPM	Measured the Moon's magnetic field.	14, 16

Lunar Ranging Retro-Reflector	LRRR	Reflected lasers fired from Earth to determine distances.	11, 14, 15
Lunar Surface Gravimeter	LSG	Measured lunar gravity.	17
Lunar Surface Magnetometer	LSM	Measured variations of the Moon's magnetic field.	12, 15, 16
Lunar Seismic Profiling Experiment	LSPE	Four geophones measured effects of eight explosive charges.	17
Neutron Probe Experiment	NPE	Measured neutron capture rates in the regolith using a 92½in-long rod (2.35m).	17
Passive Seismic Experiment	PSE	Investigated the Moon's structure. Formed part of a network with other seismometers to give precise location information through triangulation of meteorite impacts and Moonquakes.	11, 12, 14, 15, 16
Solar Wind Composition Experiment	SWC	Collected atomic particles.	11, 12, 14, 15, 16
Solar Wind Spectrometer	SWS	Measured protons and electrons from the Sun.	12,15
Suprathermal Ion Detector Experiment	SIDE	Detected positively charged ions near the lunar surface.	12, 14, 15
Surface Electrical Properties Experiment	SEP	Electromagnetic radiation measured the regolith and lunar substructure. Waves were sent by a static antenna and picked up by a receiver on the rover.	17
Portable Traverse Gravimeter Experiment	TGE	Measured differences in the value of gravity at points away from the landing site.	17

APOLLO 12

Crew:	Pete Conrad (CDR) Alan Bean (LMP) Dick Gordon (CMP)
Mission patch:	'Apollo XII' above illustration of a clipper ship above the landing site, with the astronauts' surnames below. Four stars represent the crew and original crew member C.C. Williams who died in an aircraft accident
Date (launch):	14 November 1969
Date (return):	24 November 1969
Mission duration:	10 days, 4 hours, 36 minutes
LM:	*Intrepid* (name chosen from a list submitted by Grumman employees)
CM:	*Yankee Clipper* (name chosen from a list submitted by North American employees)
Landing site:	Oceanus Procellarum (Ocean of Storms)
Time on the Moon:	1 day, 7 hours, 31 minutes
Mission objective:	To further expand lunar exploration and demonstrate accurate landing capability

The Crew

Gordon and Conrad had become friends when serving on the same Navy squadron. Conrad, who had instructed Bean as a test pilot, suggested Bean apply to be an astronaut. The crew were able to carry out the work and training required for the mission but also openly enjoy themselves at the same time.

Gold Corvettes

Although astronauts could not endorse commercial products, a Florida car dealer offered a special lease deal for astronauts. The Apollo 12 crew took up the offer – the deal cost them $1 a year – and received matching gold-painted Chevrolet Corvette coupés. Apollo 12's back-up crew (who later flew on Apollo 15) had their cars painted red, white and blue.

Missed Landing

Conrad and his crew were assigned to be back-up to Jim McDivitt's planned LM test flight. When McDivitt turned down the chance of the flight to the Moon as Apollo 8 his crew moved one position back in the flight schedule. Conrad's crew moved back with them and so lost the chance of being the first to walk on the Moon, as the rotational plan devised by Deke Slayton meant a crew would be back-up, miss two flights then fly as prime crew. Apollo 8's back-up crew was commanded by Neil Armstrong.

SCE to Aux

Just over half a minute after launch a bright flash was seen by Conrad and the CM's control panel lit up with warning lights. As the Saturn continued to fly, those on the ground tried to work out what had happened. EECOM John Aaron suggested that a switch on the CM's control panel was moved from 'SCE to Aux'. Few knew what this referred to. Bean was the only crew member who knew where the switch for SCE – the Signal Conditioning Equipment – was and when he moved it to 'Auxiliary' telemetry returned. Flight controllers were then able to diagnose the problem and issue instructions to continue the mission.

It was later determined the Saturn had been struck by lightning, the electrically charged clouds being triggered by the rocket's metallic composition and exhaust plume.

Mission controllers were concerned the parachutes needed for splashdown were damaged but reasoned nothing would be gained from aborting the mission before it had gone to the Moon and so nothing was said to the crew.

The Snowman

The landing site was among a collection of craters thought to resemble a snowman.

Pete's Parking Lot

The crew nicknamed the landing site 'Pete's Parking Lot'. It was not officially named.

'Good Godfrey'

Conrad did not want to cause offence by using profanities and used made-up euphemisms.

EVA 1

Duration: 3 hours, 56 minutes.

After collecting samples, planting the flag, and deploying the ALSEP, Conrad and Bean made a short traverse to Middle Crescent crater.

First Words

On descending the ladder Conrad said, 'Whoopie! Man, that may have been a small one for Neil, but that's a long one for me.' Conrad was one of the shortest astronauts. He had bet a journalist $500 he could say whatever he wished, but never received the money.

Lost TV

When Bean was setting up the colour TV camera it was inadvertently pointed at the Sun, burning out its sensor and so no surface television footage was seen.

Astronaut's Pin

Each astronaut was presented with a silver pin, in the shape of a shooting star, and when they flew into space they were given a gold pin. One of the first things Bean did was throw his silver pin away; it landed in a crater.

EVA 2
Duration: 3 hours, 49 minutes.

On their traverse of around 0.8 miles (1.3km) Bean and Conrad visited several craters. They collected samples, took photographs and retrieved parts of a spacecraft that had been on the Moon since 1967. The LM had accurately landed 535ft (163m) from the robotic lander Surveyor 3 and sections, including the TV camera, were removed for retrieval to Earth. On examination one part was contaminated by bacteria, suggesting it had remained active since leaving Earth and throughout its lunar stay. There has been speculation the sample became contaminated after it returned to Earth.

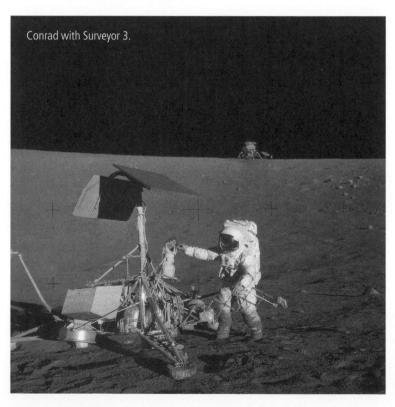

Conrad with Surveyor 3.

Space Selfie

Bean and Conrad had hidden a camera self-timer to take a photograph of themselves. Unfortunately, they were unable to find the timer in a bag filled with rocks and the chance of a unique image was gone. Bean later created the missed opportunity in one of his paintings.

55 Minutes

After lift-off and rendezvous, the LM's ascent stage was jettisoned and sent to impact the Moon. Its impact was measured by the seismometer for 55 minutes. This was a phenomenon not experienced on Earth and stunned geologists, who later explained that, unlike on Earth, there was no moisture in the lunar rocks to dampen the vibrations.

Lunar Module Pilot

LMPs did not normally operate the flight controls, however, on their way to rendezvousing with the CSM, Conrad granted Bean a chance to fly the ascent stage, when they were behind the Moon, out of contact from Mission Control.

Eclipse

On the homeward journey the crew became the first humans to see the Earth eclipse the Sun.

ΛPOLLO 13

Crew:	Jim Lovell (CDR) Fred Haise (LMP) Jack Swigert (CMP)
Mission patch:	'Apollo XIII' and the motto 'Ex Luna, Scientia' (from the Moon, Knowledge) with an illustration of the Sun and three horses representing the Greek god Apollo and his chariot, seen above the Moon
Date (launch):	11 April 1970
Date (return):	17 April 1970
Mission duration:	5 days, 22 hours, 54 minutes
LM:	*Aquarius* (the constellation and also Egyptian god who brought fertility and knowledge to the Nile region)
CM:	*Odyssey* (symbolising an epic voyage)
Landing site:	n/a
Time on the Moon:	n/a
Mission objective:	To land in the Fra Mauro Highlands area

German Measles
In the run-up to the mission, back-up crew member Charlie Duke contracted German measles and medical staff thought there was a risk the prime crew could become ill with the disease while in space. Lovell and Haise were not thought to be at risk, but Mattingly had no immunity and was replaced by Jack Swigert from the back-up crew only 72 hours before launch. Fortunately, Swigert was an expert on the workings of the CM.

Are the flowers in bloom in Houston?

Once in space, Lovell used this cryptic message to ask Capcom Vance Brand if Mattingly had started showing the spots associated with measles. He never caught the disease.

Unlucky 13?

Apollo 13 took off at 1313 hours (Houston time) and the explosion occurred on the 13th of the month. This coincidence would have been avoided had the mission launched on the intended date of 12 March. A delay was introduced to allow more time to prepare.

Pogo

Vibration during the S-II stage burn caused the central engine to shut down early but the remaining four continued to operate and the spacecraft reached orbit successfully.

Tax

Swigert brought laughter from Mission Control when he asked if he could gain an extension to submit his tax return, which he had forgotten. He was told as he was out the country it would be okay to delay submission.

Bored

> We're bored to tears down here.
>
> Capcom Joe Kerwin, 46 hours into the flight

The trans-lunar coast had been straightforward. Fifty-five hours into the flight a TV broadcast was made, which was not picked up by any of the US networks. Less than 10 minutes after it ended events were to make the mission the focus of much media and public interest.

Problem

> Houston, we've had a problem here.
>
> Jack Swigert

In the CSM liquid oxygen was mixed with liquid hydrogen in three fuel cells to produce water for drinking, system cooling and electricity. The gases were kept at very low temperatures in cryogenic tanks. Swigert was asked by Mission Control to 'stir the tanks', to allow accurate readings of their contents. The stirring was done by fans inside the tanks. Moments after Swigert began this operation, a bang was heard in the capsule. An immediate loss of power was noticed and the crew and those on the ground frantically tried to work out what had happened.

We're venting something out into space.

Thoughts it might be an instrumentation problem soon disappeared when Lovell made this observation. Both oxygen tanks were affected, with all pressure gone from Number 2 and Number 1 falling rapidly. With no oxygen there would be no power, save for that from the CM's short-term re-entry batteries.

Plume
Telescope observers on Earth could see the escaping gas. The plume expanded to become over 30 miles (48km) wide.

203,996 Miles
Apollo 13 was 203,996 miles (328,300km) from Earth when the explosion took place.

Navigation and Control
The crew found it difficult to control the spacecraft as escaping gas acted like a thruster, causing it to rotate unpredictably. It was also difficult to take any star position readings through the plume.

Lifeboat
With power and oxygen depleted in the CM, the decision was made to use the LM as a lifeboat. The LM was designed for two people to live in for 2 days and would now have to be home for three astronauts for 4 days.

Route Home

The landing was abandoned; the mission's main goal was to secure the astronauts' safe return. There were two options: 'direct abort', where the spacecraft would fire its engine, halt its movement towards the Moon and then accelerate back towards Earth, or a 'circumlunar abort', where they would continue to the Moon then swing back towards Earth. Both required an engine burn but the CSM's engine could not be risked as its condition was unknown, so the LM's descent stage engine was to be used for a circumlunar abort.

Free-Return

Apollo 13 was on a 'hybrid' trajectory, which differed from the 'free-return' type of earlier missions. A spacecraft on a free-return route would go round the Moon and slingshot back towards Earth without having to change course. Apollo 13's hybrid course, if not altered, would miss the Earth by 45,000 miles (72,420km). A course correction burn was made 5½ hours after the explosion.

Carbon Dioxide

The build-up of carbon dioxide became a problem. The lithium hydroxide canisters used to filter it out of the cabin's atmosphere were different shapes in each spacecraft: the LM's were circular and the CM's square. An ad hoc solution was devised using a plastic bag, a towel, a piece of cardboard, gaffer tape and a length of hose. The crew were carefully talked through the instructions on how to construct it. It operated successfully.

Nuclear Material

Another issue that arose was the return to Earth of nuclear material, which powered the ALSEP. The 8.5lb (3.9kg) of plutonium was enclosed in a protective case but to further alleviate any risk the LM was directed to crash in the Pacific Ocean in an area known for its deep water.

Cold
The loss of electrical power, and the need to conserve what remained, meant that heating and other systems had to be turned off, subjecting the crew to cold temperatures. Condensation formed on instrument panels and there were concerns the cold could affect electrical systems when needed for re-entry. The temperature in the CM reached 3°C (38°F) – normally it would be 22°C (72°F). The LM was a warmer 9°C (49°F).

'Upstairs Bedroom'
Swigert nicknamed the CM the 'upstairs bedroom' as it was accessed by going 'upwards' through the roof of the LM. It was rare for deep sleep to be enjoyed, with the cold and noise from communications below. The crew's lightweight garments did not provide much warmth and their sleeping bags were chiefly designed to act as restraints.

3 Hours
On average the crew received only 3 hours of sleep each night.

6 Ounces
Each astronaut drank 6oz (180ml) of water – a sixth of the normal daily intake. Water was being conserved for cooling purposes and it was thought venting of urine outside could cause course deviation. As a result of not drinking enough, Haise spent much of the return journey unwell with a kidney infection.

Frozen
When Swigert retrieved hot dogs from the CM's food storage lockers he found them frozen solid. The spacecraft had been put into a thermal rolling manoeuvre to prevent one side becoming too hot or too cold but with the internal heating off and little heat from operating equipment it wasn't enough to prevent items freezing.

PC + 2

Once the spacecraft had rounded the Moon, another burn was required. This was known as the 'pericynthion plus 2 hours' burn, with pericynthion being the point they were nearest to the lunar surface. Firing the engine sped up the return journey by 12 hours.

10 Per Cent

It was felt by some in NASA that the crew had a 10 per cent chance of survival.

Course Correction

Apollo 13 had to make a third use of the LM's engine, as the trajectory was too shallow and there was a risk of the spacecraft bouncing off the Earth's upper atmosphere and out into space. Without using the guidance computer, the crew manually lined up the spacecraft for the burn with the Sun and Earth as reference points. A further course correction was made later, as they approached Earth, using the manoeuvring thrusters.

Start-Up

The procedures that saw the CM being powered up after a prolonged period of inactivity were written in several days, when it would normally take months. There were concerns of short circuits from condensation being so close to the electrical wires but work carried out after Apollo 1 helped safeguard the circuitry.

Tiger Team

Gene Kranz was flight director on duty in Mission Control when the explosion took place. When his shift was over his White Team formed the nucleus of a 'Tiger Team' to work on producing procedures for this unprecedented situation. Astronauts and technical staff came in to help the collective effort, which left many drained and exhausted.

Apollo 13's Service Module.

Jettison

There's one whole side of that spacecraft missing.

A few hours out from Earth, the Service Module was jettisoned and the crew saw the damage for the first time. A panel had been completely blown off. The investigation suggested an explosion in the number 2 tank was caused by a short circuit from exposed wires.

Blackout
On re-entry Apollo 13 was out of contact with Mission Control for 6 minutes; communications blackout normally lasted around 3 minutes. As its trajectory was shallower, it remained out of contact for longer than usual so increasing the tension over whether the explosion had caused damage to the spacecraft's heatshield.

Rain

As they re-entered the atmosphere the crew experienced rain, not from the Earth's weather, but from condensation forming droplets inside the capsule.

14 Pounds

Lovell lost 14lb (6.4kg) in weight. The crew arrived back on Earth tired, hungry and dehydrated. The situation was worse for Haise who took three weeks to recover fully.

Attention

> Never in recorded history has a journey of such peril been watched and waited-out by almost the entire human race.
>
> *Christian Science Monitor*

As with Apollo 11, the world came together to await news. America's Space Race rival the USSR offered assistance, sending navy ships to the proposed splashdown area.

Uncertainty

> I have never experienced anything like this in my life and I never hope to experience it again.
>
> Marilyn Lovell, speaking immediately after the flight

Under the media spotlight, the crew's family endured days of uncertainty until the CM was seen on TV suspended under its parachutes on its way to splashdown.

$312,421.24

Employees of Grumman (who built the LM) sent a spoof invoice to North American Rockwell (who built the CM) for services rendered by their spacecraft. It totalled $312,421.24 and included items such as towing, battery charging and sleeping accommodation for an extra guest.

A Successful Failure

Many regarded the efforts to save the crew as NASA's finest moment. Commander Jim Lovell described it as a 'successful failure'.

Apollo 13

The dramatic events were perfect material for a Hollywood film and in 1995 Tom Hanks starred as Jim Lovell, Kevin Bacon as Jack Swigert and Bill Paxton as Fred Haise. Jim Lovell made a cameo as the US Navy admiral welcoming the astronauts home.

Failure is not an option.

Although often quoted, the line that formed the title of Gene Kranz's autobiography was not uttered during the actual flight but created by the film's scriptwriters.

APOLLO 14

Crew:	Alan Shepard (CDR) Edgar Mitchell (LMP) Stuart Roosa (CMP)
Mission patch:	'Apollo 14' and the astronauts' surnames surround the NASA astronaut pin journeying to the Moon
Date (launch):	31 January 1971
Date (return):	9 February 1971
Mission duration:	9 days, 0 hours, 2 minutes
LM:	*Antares* (star used for navigation)
CM:	*Kitty Hawk* (location of Wright brothers' first flight)
Landing site:	Fra Mauro Highlands
Time on the Moon:	1 day, 9 hours, 30 minutes
Mission objective:	Following Apollo 13's failure, to be first landing in the lunar highlands

The Three Rookies

Roosa and Mitchell hadn't flown in space before and, despite the 47-year-old Shepard being the first American in space, he was jokingly called a rookie by his fellow astronauts. Surgery enabled Shepard to return to flight status after being grounded with an inner ear problem.

Docking

In the Saturn V 'stack', the LM was positioned underneath the CSM. After TLI, the CSM detached from the S-IVB, pitched 180 degrees, then returned to dock with the LM and pull it out of its adapter. After docking the S-IVB was discarded.

On Apollo 14 the initial docking manoeuvre failed as the latches to hold the spacecraft together wouldn't lock. Attempt after attempt was made and the mission was in jeopardy. Shepard thought of pulling the spacecraft together manually but this wasn't put to the test, as finally, on the sixth docking attempt, the latches caught.

ESP

Although experiments were deployed on the Moon, an unusual one was performed in space. Mitchell had an interest in extra-sensory perception (ESP) and during quiet moments sent telepathic messages to recipients on Earth, which Mitchell claimed were successful.

Problems

The flight encountered further problems when the LM was preparing to descend to the surface. An abort light came on, which would cause the landing to be abandoned if it reoccurred on the descent. A hasty solution involving reprogramming the computer was found and radioed up, and Mitchell manually entered the changes into the computer.

During the powered descent, the LM's landing radar refused to operate properly. Switching it off and on fixed the problem and the landing continued. There was speculation about whether Shepard would have attempted the landing without radar assistance.

174ft

The LM landed 174ft (53m) from the target area. It was the most accurate Apollo landing.

EVA 1

Duration: 4 hours, 47 minutes.

In the first EVA Shepard and Mitchell took a contingency sample, deployed the ALSEP, planted the US flag and set up the TV camera.

First Words

> Al is on the surface. And it's been a long way, but we're here.
>
> Alan Shepard

Shepard was proud of finally reaching the Moon and later admitted he cried when he saw the fragility of Earth from the lunar surface.

Bands

Apollo 14 was the first mission where the Commander could be identified by red bands on his spacesuit and helmet. On previous missions it had been difficult to identify each astronaut.

EVA 2

Duration: 4 hours, 34 minutes.

'The Rickshaw'

The Modular Experiment Transporter – a two-wheeled vehicle pulled by the astronauts – was nicknamed 'The Rickshaw'. It was used to transport equipment and also acted as a workbench. On the second EVA it had to be carried as it couldn't be pulled easily on the rock-covered surface. It was only used on this mission.

Cone Crater

The astronauts aimed to reach the top of the 1,110ft-wide (338m) Cone crater. Shepard and Mitchell found that navigation was not simple with craters easily hidden and distance difficult to judge. Despite strenuous efforts up the crater's slope – Shepard's heart rate reached 150bpm at one point – with oxygen and water supplies running low, they were advised to turn back without reaching the crater's summit. It was later found they were only 65ft (20m) away from their objective. Despite not making the actual crater they were able to retrieve samples that dated the Imbrium Basin's formation. They had walked the furthest of any EVA: 9,100ft (2,800m).

Scotch
Back-up crew members Gene Cernan and Joe Engle had bet a case of whisky that Shepard and Mitchell wouldn't reach the top of the crater with the MET. While Shepard and Mitchell paid up, they were not to lose out: grateful geologists sent a case of the drink to their quarantine facility.

Sport on the Moon
At the end of the EVA Shepard produced a golf club (the head of a six-iron attached to the shaft of a sample collector) and some golf balls. The first shot ended up in a bunker (or in Moon parlance a 'crater') but the second one travelled – according to the first lunar golfer – 'miles and miles and miles', although he later admitted it landed near the ALSEP, 500ft (152m) away.

A few minutes later Mitchell threw a 'javelin' – the Solar Wind Collector's staff. It landed in the same crater as Shepard's first ball.

Charged Particles
When the LM ascent stage impacted the Moon, the ALSEP's Charged Particle Lunar Environment Experiment measured the resulting cloud of material thrown up at 8.9 miles (14km) across.

Hycon
One of the tasks assigned to Roosa in the CSM was to look for future landing sites using a Hycon Lunar Topographic Camera. Unfortunately it stopped working and he spent much time attempting to fix it before it was abandoned.

Knocking
Showing his sense of humour, Roosa responded, 'Who's there?' to Shepard's knocking on the CM hatch after rendezvous and docking.

Apollo 14 approaches splashdown.

Moon Trees

Around 500 tree seeds were taken by Roosa to see if they would survive being subjected to radiation and zero gravity. The experiment was conducted in conjunction with the US Forestry Service in which Roosa had served as a smoke jumper. A 'Moon' Sycamore was planted by Roosa's grave in Arlington Cemetery.

○○○○○○○○○○○○○○○○○○○○○○○○○○○○○○○○○

APOLLO 15

Crew:	Dave Scott (CDR) Jim Irwin (LMP) Al Worden (CMP)
Mission patch:	'Apollo 15' and the astronauts' surnames surround three bird-like symbols (representing the crew) flying over the landing site
Date (launch):	26 July 1971
Date (return):	7 August 1971
Mission duration:	12 days, 7 hours, 12 minutes
LM:	*Falcon* (mascot of the US Air Force Academy)
CM:	*Endeavour* (after Captain Cook's ship)
Landing site:	Hadley-Apennine
Time on the Moon:	2 days, 18 hours, 54 minutes
Mission objective:	To explore the Hadley-Apennine region, test lunar rover and carry out lunar orbital photography and experiments

J-Type Mission

Apollo 15 was the first of the advanced missions. Astronauts' back-packs, their spacesuits and the LM had all been upgraded to allow longer EVAs and greater payloads to be carried. The astronauts were supplied with provisions to facilitate being on the Moon for three days. The CSM also had new scientific equipment.

Trans-Lunar Issues

On the way to the Moon several issues came up. A control panel light indicating the CSM's engine was firing became illuminated. It was found to be a short circuit and a solution was radioed up. When the crew entered the LM they found shards of glass from a shattered instrument panel – a potentially dangerous issue in the weightless confines of the cabin. A vacuum cleaner and duct tape were used to collect the glass.

The King
Like other CMPs, Worden was instructed by Farouk El-Baz in orbital lunar observation and photography. The Egyptian geologist was nick-named the 'King' and Worden said while in lunar orbit, 'After the King's training, I feel like I've been here before.'

Orbit
A different lunar orbital procedure than for previous missions was used. The two spacecraft were placed in an elliptical orbit 67.3 miles (108.3km) high at one side of the Moon but only 10.6 miles (17km) at the other. This was to remove the need for the LM's Descent Orbit Insertion burn, thus saving fuel. After undocking, the CSM fired its engine to return to a circular orbit. The acceleration swung Worden's couch away from the control panel and he experienced a 'scary moment' until being able to swing it back to reach the controls again.

Landing Site
The chosen site would allow investigation of Mare Imbrium (Sea of Rains) and the Apennine Mountains – formed after the impact that created the Imbrium basin. The site was next to Hadley Rille in Palus Putredinis (Marsh of Decay) and it was hoped material from the pre-Imbrium impact lunar crust might be found.

Landing Approach
The LM had to fly a steeper landing approach (25 degrees as opposed to 14 degrees previously) to fly over the surrounding mountains, which reached 15,000ft (4,572m). When they were around 9,000ft (2,743m) above the surface, Scott and Irwin were interested to see the summit of Mount Hadley Delta thousands of feet above them – something not seen in the simulator. Scott felt they were floating by and were not going to land at their intended spot.

Targeting

Four craters called Matthew, Mark, Luke and Index were not to be seen when the LM pitched over to allow the crew to see their landing site. The LM's approach was 3,000ft (915m) too far south. Scott altered course and landed 1,800ft (550m) from the intended landing point.

Landing

> Bam!
>
> Jim Irwin

Scott hit the engine cut-off switch as soon as the Contact Light came on and the LM landed with a noticeable thump. It had the highest descent rate of all LM landings at 6.8ft (2m) per second – double that of its predecessor.

> OK Houston, the *Falcon* is on the Plain at Hadley.

Scott announced their landing with a reference to 'The Plain' – the parade ground at West Point military academy he and Worden attended.

11 Degrees

Falcon stood with one landing leg in a small crater, causing it to lean back at an 11-degree angle – 1 degree from the safety limit. This proved fortunate as when 2½ gallons (11 litres) of water leaked into the cabin, because of the incline it gathered in the back rather than near vulnerable systems at the front.

Stand-Up EVA

Soon after landing, Scott performed a lunar first: a stand-up EVA. The cabin was depressurised and he stood on the LM's ascent engine cover and opened the docking hatch. He had an unobstructed 360-degree view around Hadley Base and took photographs and made verbal observations. (Irwin later described him as 'a frustrated tank commander'.) Scott compared the view to photographs by Ansel Adams, the American landscape photographer famous for his black and white images of the American desert.

500mm
A telephoto lens was taken for the first time. It was useful in picking out features unable to be visited. The extra weight was traded off by taking less descent stage fuel.

Sleep Period
Unlike previous crews, Irwin and Scott did not venture out onto the lunar surface immediately. A sleep period was scheduled, knowing they had an intensive few days ahead. Scott and Irwin were the first on the Moon to remove their spacesuits to aid rest.

EVA 1
Duration: 6 hours, 32 minutes.
Distance travelled: 6.4 miles (10.3km).

First Words

> As I stand out here in the wonders of the unknown at Hadley, I sort of realize there's a fundamental truth to our nature. Man must explore. And this is exploration at its greatest.
>
> Dave Scott

> Boy, that front pad is really loose, isn't it!
>
> Jim Irwin

After coming down the ladder, Irwin had spun round on the footpad that wasn't fully on the surface, due to the tilt. He avoided falling by hanging onto the ladder with one hand.

Following difficulties in getting the rover deployed, Scott and Irwin set out southwards towards Elbow and St George craters on Mount Hadley Delta.

Seatbelt Basalt
While returning to the LM, Scott collected a rock by subterfuge, telling Mission Control there was a problem with his seatbelt. He was sure he'd be told to leave the rock behind if he'd mentioned it openly.

Heat Flow
Back at the LM, Scott and Irwin deployed the ALSEP. One experiment caused problems: the holes for the heat flow experiment proved difficult to create using the battery-powered drill. Despite strenuous efforts, it was not possible to reach the intended depth of 120in (3m), with 70in (178cm) reached.

Temperature Change
The ALSEP operated during a lunar eclipse. As the Sun's light disappeared, the equipment measured a temperature decrease rate of 127°C (260°F) per hour.

Lunar Rover
The rover provided a huge step in capability, although safety considerations meant it could only be driven a distance that the astronauts could walk back if it suffered a malfunction. In the weak lunar gravity the rover tended to lift off – up to a foot (0.3m) in height – when it hit any bumps; all four wheels were observed to have left the surface at one point. It could also lose traction and skid when turning at speeds over 6mph (9.7km/h).

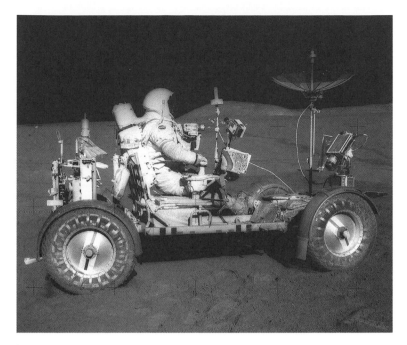

Lunar rover.

TV
For the first time, the TV camera was remotely controlled from Earth. EVAs were broadcast live, as was the ascent stage's lift off, the camera being mounted on the rover.

EVA 2
Duration: 7 hours, 12 minutes.
Distance travelled: 7.8 miles (12.5km).

The rover was again driven south, towards the Apennine Mountain Front at Mount Hadley Delta. Spur and Dune craters and several other locations were visited.

15 Degrees

Mount Hadley Delta's slope was 15 degrees and Scott and Irwin found it difficult to walk uphill. It also presented difficulties for the rover, which would slide downhill if not physically held by the astronauts.

Genesis Rock

> I think we found what we came for.
>
> Dave Scott

Sample number 15415 became known as the Genesis Rock. Found at Spur crater, this white anorthosite rock was 4.1 billion years old, dating to before the Imbrium impact. Scott and Irwin thought the rock was almost presented to them, waiting to be found, sitting on a pedestal of lunar soil.

Green Glass

Scott and Irwin made another significant discovery at Spur when a sample of green material was taken. It was thought to be caused from a volcanic eruption, which caused the material to form beads. The colouring was caused by a high proportion of magnesium.

Core Sample

A core sample was to be taken, which required a 10ft-deep (3m) hole to be drilled. The same drill as for the heat flow experiment was used but it struggled to dig through compacted regolith and Scott suffered sore hands and a sore shoulder from the effort. (Irwin had also experienced pain in his fingers and cut his nails to alleviate the discomfort within the tight-fitting gloves). Eventually a 7.9ft-long (2.4m) sample was recovered. When analysed back on Earth it was found to contain forty-two different layers.

○○○○○○○○○○○○○○○○○○○○○○○○○○○○○○

EVA 3
Duration: 4 hours, 49 minutes.
Distance travelled: 3.1 miles (5.1km).

The last EVA began late and because of prioritising experiments its duration was curtailed. The crew visited Hadley Rille to the west, but were not able to go to the North Complex. Lift-off was scheduled for the same day and time to prepare was required.

Hadley Rille
Hadley Rille is a collapsed lava tube running for 80 miles (129km) across the lunar surface. It is a mile (1.6km) across and 1,000ft (305m) deep. As the astronauts examined the rille, there were concerns from those in Mission Control watching on TV they might fall in.

Galileo
Before ending the EVA, Scott had a surprise. He produced a falcon feather and held it out with one hand. With the other he held out a geology hammer. He then let both go at the same time. According to Galileo, they would fall at the same rate, as gravity affects objects of different mass with the same force. On Earth air resistance affects descent rates but on the Moon they both hit the surface simultaneously.

Capcom Joe Allen later wrote in a report that this was reassuring as 'the homeward journey was based critically on the validity of the particular theory being tested'.

Postal Service
Another activity carried out before leaving was the cancellation of a stamp. Eight-cent stamps marking the space programme had been issued by the US Postal Service and two were taken to the Moon on an envelope. Scott used a rubber stamp to put the postmark 'UNITED STATES / ON THE MOON / AUG. 2 1971 / FIRST DAY OF ISSUE' on this unique first day cover.

Higher Power

'I look unto the hills, from whence cometh my help.' But of course, we get quite a bit from Houston, too.

Jim Irwin

Irwin quoted from the Bible while looking at the mountains at the end of EVA 3. Scott left a small Bible on the rover.

Fallen Astronaut

Before leaving the lunar surface for the final time, Scott left a small memorial to astronauts and cosmonauts who had died. The memorial took the form of a plaque and a small aluminium statue called 'Fallen Astronaut', by Belgian artist Paul van Hoeydonck.

The event was later the source of some controversy when plans to make and sell replica statues were announced by the artist's gallery, something not welcomed by NASA who had to deal with the fallout from a German dealer selling first day covers brought back from the Moon by the crew. They had taken the covers on the understanding they wouldn't be sold until after Apollo had finished. They were reprimanded and it cast a shadow over the achievements of the mission.

Heart Problems

Heart irregularities were noticed in both Scott and Irwin by medical staff but they were not informed of the situation. The amount of exertion and a low level of potassium were put down as the cause.

Irish Flag

Irwin, who had Irish ancestry and was born on St Patrick's Day, took an Irish flag to the Moon, which he presented to the country's president in 1979. For the inscription he wrote, 'Jim Irwin found Green Rock near "Genesis" Rock. Significant for Irishman to find such a rock!'

Irwin collects a sample.

○○○○○○○○○○○○○○○○○○○○○○○○○○○○○○○○○

SIM Bay

The SIM (Scientific Instrument Module) bay was installed in a section of the Service Module. It contained mapping and panoramic cameras, sensing equipment and a small 84lb (38kg) sub-satellite, left in lunar orbit to measure charged particles, lunar gravity and magnetic fields. On Apollos 15 and 16, mass and gamma ray spectrometers were mounted on booms 24 and 25ft-long (7.3m and 7.6m) respectively. They had to be retracted before the engine burn to leave lunar orbit or they would snap off.

6,500ft
Each negative for the panoramic camera measured 45.24in x 4.5in (115cm x 11.4cm). It had capacity for 1,650 exposures; the film was 6,500ft (almost 2km) long.

Worden
Al Worden appears twice in the *Guinness Book of Records.* He is recognised as the most isolated human being as during Apollo 15 he was 2,235 miles (3,597km) from his crewmates and spent 73 hours on his own. The other award was for the first deep-space EVA when he recovered film cassettes from the SIM bay while almost 200,000 miles (321,869km) from Earth. He spent 38 minutes outside and was afforded a unique view of Earth and the Moon.

Cinder Cones
From orbit Worden noticed a field of cinder cones in the Taurus-Littrow area. This led to its selection as Apollo 17's landing site.

Into the Wild, Blue Yonder
As the ascent stage lifted off the airwaves were suddenly filled with the US Air Force song being played at loud volume. The music came from the CM. Worden had intended it to be heard only in Mission Control and blamed them for patching it through to the LM during the critical part of their flight. Irwin later said that he was meant to play it after a minute not a few seconds.

Two Chutes
The return journey to Earth passed uneventfully but when the CM was 6,000ft (1,829m) above the sea on its parachute-assisted descent one of the parachutes was seen to deflate. The CM hit the water harder and sooner than expected. It was thought the dumping of the CM's monomethyl hydrazine thruster fuel had burnt holes in the nylon parachute and its rigging.

170 Pounds

Apollo 15 returned 170lb (77kg) of rock and soil samples, equivalent to those from Apollo 12 and 14 combined.

APOLLO 16

Crew:	John Young (CDR) Charlie Duke (LMP) Ken Mattingly (CMP)
Mission patch:	Sixteen stars and the astronauts' surnames surround an eagle on top of a blue, red and white shield with 'Apollo 16' on the top and the lunar surface behind. A yellow vector derived from the NASA logo is placed in the centre
Date (launch):	16 April 1972
Date (return):	27 April 1972
Mission duration:	11 days, 1 hour, 51 minutes
LM:	*Orion* (the constellation)
CM:	*Casper* (the astronaut's white garments were said to resemble Casper the Friendly Ghost)
Landing site:	Descartes Highlands
Time on the Moon:	2 days, 23 hours, 2 minutes
Mission objective:	To explore lunar highlands; carry out orbital photography and experiments

Dream

Six months before the mission, Duke had a dream in which he and Young were driving on the Moon when they came across another rover. When he lifted the visors of the seated astronauts he saw himself and Young, both dead.

Twin

Duke's twin brother came to Florida to watch the launch. When he was seen by NASA management at a hotel, they assumed it was Charlie, breaking the quarantine imposed before launch.

70

At lift-off Young's heart rate was 70bpm; Duke's was 144. Young said it was because he was older and his heart wouldn't go any faster.

Problems

Launch and TLI were without major incident; however, 38 hours into the flight, the crew received a warning the IMU had gone into gimbal lock. The platform would have to be realigned but the stars could not be sighted due to broken-off paint fragments from the LM obscuring the view. Instead it was done by using the positions of the Moon and Sun. A computer program to avoid the issue reoccurring was radioed up by Mission Control.

> It's really the worse sim I've ever been in.
>
> John Young

Before descent the LM crew experienced problems with the S-Band antenna and pressure of the reaction control thrusters.

> I be a sorry bird.
>
> Ken Mattingly

After undocking before PDI, Mattingly prepared to fire the CSM's engine to move into a circular orbit. However, a problem was noticed with the secondary system used for the engine. Mission rules stated if it was not fixed then the mission would be aborted. The crew were resigned to coming home early. The LM and CSM remained in close formation for almost 6 hours before the decision was taken to continue the mission.

Landing Site

The lunar highlands, which cover more than 80 per cent of the Moon's surface, had not been explored before. Geologists had hoped the mission would visit Tycho crater but this was ruled out as being too hazardous. The Descartes area (situated near to the Descartes crater) was thought to be a place where evidence of volcanic activity could be found. Two geological units would be investigated from the landing site: Cayley Formation (smooth, filled basins) and the Descartes Formation (irregular shaped, hilly topography).

EVA 1

Duration: 7 hours, 11 minutes.
Distance travelled: 2.6 miles (4.2km).

First Words

> There you are, our mysterious and unknown Descartes highland plains. Apollo 16 is gonna change your image.
>
> John Young

Once on the lunar surface, the normally reserved Young raised both arms in celebration. He then enigmatically referred to Brer Rabbit, back in 'the briar patch where he belongs', referring to the Uncle Remus story about the clever rabbit that outwits his nemesis Brer Fox.

> Fantastic! Oh, that first foot on the lunar surface is super, Tony!
>
> Charlie Duke to Capcom Tony England

A naturally enthusiastic person, Duke was noted for his exuberance while on the Moon.

At the start of the EVA, the US flag was raised, a deep core sample taken and the ALSEP deployed.

Heat Flow Experiment
An improved drill avoided the problems experienced on Apollo 15. However, when Young caught his foot on the electrical cable he caused it to be inoperative for the rest of the mission.

Far Ultraviolet Camera/Spectroscope
The first observatory to be used on the Moon used a 3in (7.6cm) telescope to take photographic images and also spectroscopic data in the far ultraviolet spectrum. It studied the Earth's atmosphere, solar wind, Milky Way and other areas.

First Traverse
Using the rover, the first objective was west towards Flag crater. Nearby was Plum crater where they stopped to collect samples. The rocks being found were breccias – indicating impact origin and not volcanism. On the return leg they stopped at Buster and Spook craters. Plum was named by Young after his daughter's nickname 'Sugarplum' and Buster after his son John, who he called 'Buster Brown'.

Big Muley
While the astronauts were at Plum crater geologists in Houston had spotted a large rock and, thinking it to be volcanic, asked for it to be collected. It was the largest rock from all the Apollo missions, weighing 26lb (11.7kg) and was named after Bill Muehlberger, Apollo 16's geology lead.

Driving Conditions

Man it all looks the same doesn't it.

Charlie Duke

John Young salutes mid-jump.

Navigation was problematic driving down-Sun, as the Sun's light reflected off the surface, making features difficult to see. Another obstacle was that landmarks on maps created from orbital photography were not always easy to see.

Grand Prix
The Grand Prix was a demonstration of the rover, filmed by Duke. At one point he told Young to turn sharply to which the driver replied, 'I have no desire to turn sharp.'

Space Shuttle

> The country needs that Shuttle mighty bad.
>
> John Young

While on the surface, news was passed that Congress had approved budget for the Space Shuttle. This would have great relevance for Young, who would command the first flight nine years later.

Swan Lager

Before they left the LM for the second EVA, a communications break-
down between Houston and the ground station at Honeysuckle Creek in
Australia led to John Saxon, Honeysuckle's operations supervisor, speaking
to the astronauts. The three discussed the attractions of being able to have
a cold beer and Saxon invited them to come over any time. When Young
and Duke returned to Earth they found cases of the brand of lager they'd
discussed had been delivered to their homes by the brewery company.

EVA 2

Duration: 7 hours, 23 minutes.
Distance travelled: 7 miles (11.3km).

The second EVA was southwards, towards the Cinco craters, located on
the slope of Stone Mountain, before heading north-west to Stubby and
Wreck craters.

500ft

Young and Duke reached a height of 500ft (153m) on Stone Mountain
where they were afforded a view described by Duke as 'spectacular'. It
was the highest elevation reached by any Apollo crew on the lunar surface. Young avoided the problems experienced on the previous mission,
when the rover slipped downhill, by parking it in a small crater.

Fender

Young accidently knocked off part of the right rear wheel's fibreglass
fender with his geology hammer, leading to them being covered in
more dust than normal. Although provided with a brush to remove it
from their suits, much lunar dust made its way into the LM.

EVA 3
Duration: 5 hours, 40 minutes.
Distance travelled: 7 miles (11.3km).

Following the delayed landing, Mission Control considered cancelling the third EVA but instead a curtailed version was undertaken. Young and Duke headed northwards, towards North Ray crater. Time was critical due to the scheduled lift-off for rendezvous with the CSM later the same day.

House and Shadow Rock
They encountered two large boulders. House Rock, described by Young as a 'biggie', was 40ft (12m) high and 80ft (24m) across. Shadow Rock was named because it overhung soil that received no sunlight. When Young investigated this area, Duke warned him, 'You do that in west Texas and you get a rattlesnake.'

Moon Olympics
With a few minutes available at the end, the astronauts started their own Moon Olympics. During the 'high jump', Duke fell backwards, landing on his backpack. He experienced several moments of panic, waiting to see if his suit had punctured.

Gas

> I like an occasional orange, really do. But I'll be darned if I'm going to be buried in oranges.
>
> John Young

Following the heart irregularities experienced on Apollo 15, Apollo 16's crew were instructed to drink orange juice to maintain levels of potassium. Young complained, on an inadvertent live microphone link, about how the juice was giving him stomach gas. Orange juice also caused Duke problems: during an EVA his drink bag leaked and smeared his face with the sticky liquid.

81 Hours, 40 Minutes

Mattingly spent 81 hours, 40 minutes on his own – a record for solo spaceflight. Like other CMPs, he had a busy schedule which became problematic when Mission Control made changes to the flight plan. At times he was forced to eat while going to the toilet. One of his tasks was taking images of the solar corona and in order not to lose his night vision, he recorded his checklist on a tape recorder to be played back.

Wedding Ring

During Mattingly's SIM bay EVA on the return journey, Duke spotted Mattingly's wedding ring – which had been missing since near the start of the mission – floating out the hatch. It proceeded to hit Mattingly, then return towards the hatch where Duke quickly grabbed it.

Family

This is the family of astronaut Charlie Duke from planet Earth who landed on the Moon on the twentieth of April 1972.

Duke left a family photograph on the surface. He later reckoned it would have faded, after being exposed to sunlight for years.

Explosion

After Apollo 15's parachute deflation, excess propellant was retained on the CM after landing. During the spacecraft's deactivation a tank cart containing thruster fuel exploded. Windows were shattered, a hole was blown in the building's roof and forty-six personnel were taken to hospital, suspected of inhaling the toxic fumes.

Mission Findings

> Well it's back to the drawing boards, or wherever geologists go.
>
> Ken Mattingly

The mission had not produced what geologists had expected. Volcanic rocks had not been found, but the findings would be used in developing alternative theories on the Moon's origin.

ΛPOLLO 17

Crew:	Gene Cernan (CDR) Harrison 'Jack' Schmitt (LMP) Ron Evans (CMP)
Mission patch:	'Apollo XVII' and the astronauts' surnames surround the Sun god Apollo, an American eagle, the Moon, Saturn and a spiral galaxy
Date (launch):	7 December 1972
Date (return):	19 December 1972
Mission duration:	12 days, 13 hours, 51 minutes
LM:	*Challenger* (representing the challenges ahead)
CM:	*America* (a tribute to the American people)
Landing site:	Taurus-Littrow Valley

Time on the Moon:	3 days, 2 hours, 59 minutes
Mission objective:	To continue extended lunar explorations, with the first ever geologist crew member

Black September
The terrorist group had killed Israeli athletes at the Munich Olympic Games earlier in the year and US intelligence received indications of threats against the astronauts and their families. Increased round-the-clock security was provided.

Injuries
Six weeks before launch, Cernan hurt his leg playing softball and was concerned over his prospects of flying. His painful tendon troubled him throughout his time on the Moon. While a member of Apollo 14's back-up crew he had crashed while flying a helicopter and could have jeopardised his chances of commanding Apollo 17.

Night Launch

It's lighting up the area, it's just like daylight here at the Kennedy Space Center.
Charles 'Chuck' Hollinshead, Public Affairs Officer

Apollo 17 was the programme's only night launch. Lift-off took place after midnight, following a delay in the countdown. The awesome sight and noise saw fish leap out of nearby waters. The Saturn's ascent could be seen 500 miles away up and down the US east coast.

It was estimated up to a million came to see Apollo's last Moon flight, and a record 42,000 official guests attended. Amongst the invited guests was Charlie Smith, who claimed to be the oldest man in America at 130 years old. (He was later found to be 'only' 98 in 1972.) Smith was sceptical about the lunar landings saying, 'If they brought back rocks, they took 'em with them.'

Scissors

En route to the Moon Evans lost his scissors and was given a pair by his crewmates. Although apparently trivial, they were essential, being used to open food packets.

Landing Site

Much deliberation went into choosing the final Apollo landing site. Schmitt had advocated landing on the far side's Tsiolkovksy crater but this was rejected in favour of the Taurus-Littrow valley. It was in the Taurus Mountains, on the south-eastern edge of the Mare Serenitatis (Sea of Serenity), 17 miles (27km) from Littrow crater. The valley offered the opportunity of retrieving samples of older lunar crust rocks from the pre-Imbrium impact period and of recent volcanism. Samples of mare material would also determine its age.

Landing

The LM flew in over the Sculptured Hills before landing on the valley floor. The canyon was no more than 5 miles (8km) across but precision landing was a matter of course by this stage in the programme.

Craters

Craters were named by the crew either to mark well-known figures or family members. They included:

Name	Origin
Barjean	Barbara Jean (Cernan's wife).
Brontë	Novelist Charlotte Brontë.
Camelot	King Arthur's castle.
Cochise	Native American chief.
Faust	Character of mythic German legend.
Gatsby	Character in F. Scott Fitzgerald's novel.

Horatio	Character in Horatio Hornblower novels.
Jones	John Paul Jones, Scottish creator of US Navy.
Lara	Character in *Dr Zhivago*.
Lewis and Clark	US explorers.
Nansen	Norwegian explorer.
Poppie	Cernan's father.
Punk	Cernan's nickname for daughter Tracy.
Rogers	Sci-fi character Buck Rogers and US comedian Will Rogers.
Rudolph	Mission was close to Christmas.
Shakespeare	English playwright.
Sherlock	Character in Arthur Conan Doyle stories.
Shorty	Character in a novel called *Trout Fishing in America*.
Spirit	Charles Lindbergh's *Spirit of St Louis* aircraft.

OOOOOOOOOOOOOOOOOOOOOOOOOOOOOOOOO

EVΛ 1
Duration: 7 hours, 12 minutes.
Distance travelled: 2.2 miles (3.5km).

The first EVA was shortened due to problems setting up the ALSEP and just one short geological traverse to Steno crater was made.

First Words

> And, Houston, as I step off at the surface at Taurus-Littrow, we'd like to dedicate the first step of Apollo 17 to all those who made it possible.
>
> Gene Cernan

Singing

I was strolling on the Moon one day.

As he collected samples and took photographs, Schmitt would burst into song. He and Cernan both sang an adapted version of an old song, 'While Strolling Through the Park One Day', changing the lyrics from the 'merry, merry month of May' to 'December'. They adapted so well to working and manoeuvring on the Moon they were even seen 'skiing' downhill.

Drilling
For the Heat Flow Experiment none of the problems of Apollo 15 and 16 were experienced, and drilling for the two probes was successful, achieving a depth of 93in (2.36m). For a deep drill core sample, extraction of the 114½in-long core (2.91m) caused problems but all three sections were removed successfully.

Flag
As on previous missions a US flag was planted. Apollo 17's had hung in Mission Control throughout the Apollo programme before being taken on the flight. A replacement was flown to the Moon and on its return presented to Gene Kranz.

EVA 2
Duration: 7 hours, 37 minutes.
Distance travelled: 12.7 miles (20.4km).

Running Repairs
During the first EVA, Cernan accidently damaged the rover's right rear fender with his geology hammer. It caused the astronauts to be covered by dust when travelling. Mission Control worked overnight then radioed

up a solution: four map pages were taped together and attached with metal clamps.

Longest Drive

The EVA's main objective was to visit the 7,500ft-high (2,286m) South Massif. The drive of over an hour took Cernan and Schmitt the furthest from the landing site of any in Apollo: 4.9 miles (7.4km).

Amongst the locations they visited were three craters:

Nansen

They spent over an hour here and collected one of the oldest samples returned, a rock fragment dating 4.5 billion years old.

Lara

While there Schmitt performed gyrations while collecting samples that led to a nearby crater being named Ballet.

Shorty

While at Shorty, Schmitt observed orange soil. It was initially thought to be from volcanic activity but was found to be glass spheres, with high levels of titanium and iron, formed by fire fountains: molten lava being propelled high above the lunar surface. Cernan and Schmitt also stopped at a wrinkle ridge called Lincoln-Lee Scarp, a type of lunar feature not investigated before on any mission.

Rover Samples

Using a scoop, Schmitt was able to collect material without leaving his seat. One of the samples was from a landslide, thought to be caused by ejecta material from the impact that created Tycho crater 1,300 miles (2,093km) away.

EVA 3
Duration: 7 hours, 15 minutes.
Distance travelled: 7.5 miles (12.1km).

The astronauts drove north then east, their prime objective to sample a large boulder – 59ft (18m) in length and big enough to be seen from lunar orbit – on the slopes of North Massif. The boulder had rolled down the slope 20 million years ago and then split into five sections. Sculptured Hills and Van Serg crater were also visited.

Hammer Thrower
With no further need at the end of the EVA, Schmitt threw away their geology hammer, its flight taking it 144ft (44m) away from the LM.

TDC
After parking the rover, Cernan scratched the initials of his daughter Tracy Dawn Cernan into the lunar soil.

Schmitt next to a large boulder on EVA 3.

Final Words
Before he ascended the ladder for the last time, Cernan made a short speech. It ended:

> And as we leave the Moon at Taurus-Littrow, we leave as we came and, God willing, as we shall return, with peace and hope for all mankind.

TV Remote
As on the previous two missions, the TV camera mounted on the rover was remotely controlled from Earth. As the ascent stage launched the camera movement was timed perfectly – despite the time lag between the Earth and Moon – capturing the lift-off and initial climb.

Ron Evans

> Hello, mom. Hello, Jan. Hi, Jon. How are you doing? Hi, Jaime!

On the return journey Evans' SIM bay EVA lasted just over an hour. During the spacewalk he enthusiastically greeted his family.

Evans was the only Apollo astronaut who saw service in Vietnam. He flew F-8 fighters off USS *Ticonderoga* and was on board when he heard his application to be an astronaut had been accepted. Evans would be reacquainted with the aircraft carrier as it was the ship that recovered the crew after their mission.

Captain America
Due to his patriotism, Evans was nicknamed after the Marvel comic book character.

Orbital Experiments

In addition to the panoramic and mapping cameras and laser altimeter, the CSM carried several experiments and equipment:

Experiment	Details
SIM Bay	
Infrared scanning radiometer	Mapped the Moon's thermal emissions.
UV spectrometer	Measured lunar atmospheric density and composition.
Lunar sounder	Used radar to measure sub-surface structure.
S-band transponder	Provided accurate measurements of CSM speed, thereby measuring gravity at different locations around the Moon.
Gamma ray spectrometer	A sodium iodide scintillation crystal, brought back to Earth for analysis.
CM	
Window meteoroid experiment	Recorded impact of micro-meteorites on the CM's windows.
Biostack	Studied how cosmic rays affected biological specimens such as bacteria, beetle and grasshopper eggs, seeds and shrimp.
Biocore	Measured effects of cosmic ray radiation on the brains and eyes of five Pocket Mice, four of which survived the journey.

243

Apollo 17 brought back 243lb (110kg) of lunar samples – the most of any mission.

The Blue Marble.

Blue Marble

Apollo produced many iconic images but one of the most reproduced is called 'Blue Marble'. It is of a full Earth, taken during Apollo 17's outward journey.

○○○○○○○○○○○○○○○○○○○○○○○○○○○○○○○○

Apollo Facts and Figures

5.3

At its peak in 1965, NASA received 5.3 per cent of the US federal budget.

9
Nine manned lunar missions were flown.

700
Apart from the experiments and commemorative items, astronauts left over 700 items on the lunar surface, such as cameras, backpacks and even nail clippers.

842
Astronauts brought back 842lb (382kg) of lunar material. Some samples are kept sealed to be studied by methods developed in the future.

20,000
Over 20,000 companies and universities were involved.

30,996
During the missions, 30,996 photographs were taken.

400,000
Four hundred thousand worked on the project. The vast majority were employed by private sector contractors. Of the 23,267 workers at KSC in 1969 only 12 per cent were NASA employees.

18,009,891
Apollo spacecraft travelled over 18 million miles (28,984,109km).

$25.4 billion
Apollo cost $25.4 billion. Throughout the programme there were criticisms the money could be better spent elsewhere and a protest was held at KSC by civil rights campaigner Reverend Ralph Abarnathy on Apollo 11's launch day.

7

AFTER APOLLO

Following Apollo, NASA continued manned spaceflight operations, albeit at a different level of funding and endeavour than before.

Apollo 18, 19 & 20

Three planned missions were cancelled in 1970. The Saturn Vs assigned to these missions were used for Skylab, the Apollo–Soyuz flight and as ground display exhibits.

Skylab

In the mid sixties it was envisaged that after the lunar landing missions were completed, the Apollo Applications Program would follow. Amongst the proposals were more advanced missions with lunar bases featuring astronauts travelling by surface and flying vehicles. However, these plans were dropped as a result of budget cuts. Instead, Earth-orbital missions were funded and given the name Skylab.

In 1973, three manned missions were flown to America's first space station, made from a converted S-IVB stage. Apollo 12 astronauts Alan Bean and Pete Conrad commanded two missions. The first space repair was made when aerodynamic forces jammed one of Skylab's solar panels and tore the other off during launch. The jammed panel was released by astronaut Paul Weitz leaning out of the CM's hatch with a 10ft-long (3m) pole while having his legs held by Joe Kerwin.

In total Skylab astronauts spent 171 days in space, carrying out observations and experiments, particularly of the Sun.

There were plans for the Space Shuttle to boost it to a higher orbit but friction with the upper atmosphere caused Skylab's orbit to deteriorate and it came down in the Indian Ocean and part of Western Australia in July 1979.

Λpollo–Soyuz

On 15 July 1975 two spacecraft lifted off within hours of each other: one from Baikonur Cosmodrome in Kazakhstan, the other from KSC in Florida. Two days later Soyuz and Apollo spacecraft rendezvoused and then docked, 142 miles (229km) above the Earth.

It was a historic moment and signified the end of the Space Race. When the hatches were opened cosmonauts Alexei Leonov and Valery Kubasov shook hands with astronauts Deke Slayton, Tom Stafford and Vance Brand. Slayton had waited 16 years for his space flight. Showing spaceflight had lost none of its dangers, toxic thruster propellant entered the cabin during the parachute descent, causing the crew to be hospitalised.

Return to the Moon

Unmanned lunar missions continued until 1976, followed by a 14-year hiatus. Since the 1990s the Moon has been visited by spacecraft from Japan, China, Europe, India and the USA. Plans exist for future crewed landing missions but the last human exploration remains that of Apollo 17 in 1972.

Legacy

> You can see the Earth like a beautiful, fragile Christmas tree ornament hanging against the blackness of space.
>
> Jim Irwin

The Apollo missions remain one of the greatest examples of human ingenuity, determination and aptitude applied to one goal. However, its legacy was something no one expected: by journeying to the Moon the astronauts turned attention back on the 'good Earth'. Their photographs showing the planet as a vulnerable object hanging in the vast emptiness of space changed how those on Earth viewed their home planet.

Moon Landing Hoax

I have the grey hairs and so does Charlie Duke that prove we went!

John Young

Despite the overwhelming evidence of photographs, film, testimonies, returned physical objects and scientific experimental data, some believe the landings were faked and a huge cover-up was organised to include the 400,000 people who worked on the project.

SOURCES

Websites
NASA provides an enormous amount of material online and this book would not be possible without these sources, too numerous to mention individually. Other informative sites:

Apollo Lunar Surface Journal
www.hq.nasa.gov/alsj/main.html

Apollo Lunar Flight Journal
history.nasa.gov/afj/

CollectSpace
www.collectspace.com

Smithsonian National Air and Space Museum
airandspace.si.edu/

Space
www.space.com

US Geological Service
astrogeology.usgs.gov

Gazetteer of Planetary Nomenclature
planetarynames.wr.usgs.gov

Books

Aldrin, Colonel Edwin E. 'Buzz' and Warga, Wayne, *Return to Earth* (Random House, 1973)

Carlowicz, Michael J., *The Moon* (Harry N. Abrams, 2007)

Cernan, Eugene and Davis, Don, *The Last Man on the Moon* (St Martin's Griffin, 1999)

Chaikin, Andrew, *A Man on the Moon: The Voyages of the Apollo Astronauts* (Penguin Books, 2009)

Collins, Michael, *Carrying the Fire: An Astronaut's Journeys* (Cooper Square Press, 2001)

Collins, Michael, *Liftoff: The Story of America's Adventure in Space* (Aurum Press, 1989)

Farmer, Gene and Hamblin, Dora Jane, *First on the Moon: A Voyage with Neil Armstrong, Michael Collins, Edwin E. Aldrin Jr* (Michael Joseph, 1970)

French, Francis and Burgess, Colin, *In the Shadow of the Moon: A Challenging Journey to Tranquility, 1965–1969* (University of Nebraska Press, 2007)

Hansen, James R., *First Man: The Life of Neil Armstrong* (Simon & Schuster, 2005)

Irwin, James B. and Emerson Jr, William A., *To Rule the Night: The Discovery Voyage of Astronaut Jim Irwin* (A.J. Holman, 1973)

Kraft, Chris, *Flight: My Life in Mission Control* (Plume, 2002)

Kranz, Gene, *Failure is Not an Option: Mission Control from Mercury to Apollo 13 and Beyond* (Berkley Books, 2001)

Lindsay, Hamish, *Tracking Apollo to the Moon* (Springer, 2001)

Lovell, Jim and Kluger, Jeffrey, *Lost Moon: The Perilous Voyage of Apollo 13* (Houghton Mifflin, 1994)

Parry, Dan, *Moonshot: The Inside Story of Mankind's Greatest Adventure* (Ebury Press, 2009)

Reynolds, David West, *Kennedy Space Center: Gateway to Space* (Firefly Books, 2006)

Scott, David and Leonov, Alexei with Toomey, Christine, *Two Sides of the Moon: Our Story of the Cold War Space Race* (Simon & Schuster, 2004)

Scott, David Meerman and Jurek, Richard, *Marketing the Moon: The Selling of the Apollo Lunar Program* (The MIT Press, 2014)

Slayton, Donald K. and Cassutt, Michael, *Deke! US Manned Space: From Mercury to the Shuttle* (Tom Doherty Associates, 1994)

Smith, Andrew, *Moon Dust: In Search of the Men Who Fell to Earth* (Bloomsbury, 2005)

Stafford, Thomas P. and Cassutt, Michael, *We Have Capture* (Smithsonian Books, 2002)

Stroud, Rick, *The Book of the Moon* (Doubleday, 2009)

Ward, Jonathan H., *Rocket Ranch: The Nuts and Bolts of the Apollo Moon Program at Kennedy Space Center* (Springer Praxis, 2015)

Wilhelms, Don, *The Geologic History of the Moon* (US Geological Survey Professional Paper 1348, 1987)

Wood, W. David, *How Apollo Flew to the Moon* (SpringerPraxis, 2011)

Worden, Al with French, Francis, *Falling to Earth: An Apollo 15 Astronaut's Journey to Earth* (Smithsonian Books, 2011)

Young, Anthony, *Lunar and Planetary Rovers: The Wheels of Apollo and the Quest for Mars* (Springer Science & Business Media, 2007)

Young, Anthony, *The Apollo Lunar Samples: Collection Analysis and Results* (Praxis, 2017)

Young, John W. with Hansen, James R., *Forever Young: A Life of Adventure in Air and Space* (University Press of Florida, 2012)

TV

Neil Armstrong: First Man on the Moon (BBC, 2012)

ACKNOWLEDGEMENTS

Huge thanks go to Chrissy McMorris of The History Press for her support of the project. Thanks are also due to everyone else involved in making this book a reality, particularly Alex Waite and Jemma Cox.

Bert Ulrich of NASA and Kim Mesquita and Brooke Parsons of Apollo High School, Glendale, Arizona, are also thanked for their assistance.

The biggest thanks go to my family for allowing me the time and space to write this book.

The destination for history
www.thehistorypress.co.uk